Maple V fo Engineers

Douglas B. Meade
University of South Carolina
Mathematics Department

Etan Bourkoff
University of South Carolina
Department of Electrical and Computer Engineering

Maple V for Engineers

ADDISON-WESLEY

An imprint of Addison Wesley Longman, Inc.

Menlo Park, California · Reading, Massachusetts · Harlow, England
Berkeley, California · Don Mills, Ontario · Sydney · Bonn · Amsterdam · Tokyo · Mexico City

DEDICATION

To my parents, Kenneth and Catherine Meade

To the memory of my mother, Theresia Roth Bourkoff
and my sister, Talma Bourkoff Mitty

ACKNOWLEDGEMENTS

The authors gratefully acknowledge the assistance of Dr. Todd Jay Mitty of
Princeton University and Professors Stephen McAnally, Michael Sutton, and
John VanZee of the University of South Carolina for their contributions
to the Applications in this book.

. . . .

Associate Editor: Nate McFadden
Developmental Editor: Judy Ziajka
Production: Emilie Bauer
Copyeditor: Robert Fiske
Proofreader: Holly McLean Aldis
Indexer: Nancy Kopper
Cover Design: Yvo Riezebos
Text Design: Side by Side Studios
Composition: Craig W. Johnson

Library of Congress catalog card number 95-131339

This is a module in the *Engineer's Toolkit* edition. Contact your sales representative
for more information.

The Engineer's Toolkit is a trademark of Addison Wesley Longman, Inc.

Photo Credits for chapters 1–5: ©1997 Photo Disc, Inc.

ISBN: 0-8053-6445-5

1 2 3 4 5 6 7 8 9 10—CRK—01 00 99 98 97

Addison-Wesley Longman, Inc.
2725 Sand Hill Road
Menlo Park, CA 94025
http://www.aw.com/cseng/toolkit/

Contents

Other than their introductions, Chapters 6 and 7 of this module are not bound in this book. You can access them on the Addison Wesley Engineer's Toolkit website at <http://www.awl.com/cseng/toolkit/modules/map>.

1 Problem Solving with Maple

Measuring Distances with Lasers One of the ten great engineering achievements identified by the U.S. Academy of Engineering is light amplification by stimulated emission of radiation—more commonly known as the laser. The use of lasers to measure distances, large and small, has become routine; for example, civil engineers regularly use lasers in surveying, and NASA engineers use lasers to measure the distance between objects in space. Accurate measurements of the time a pulsed laser beam takes to travel from earth, reflect off the moon, and return to an optical receiver at the laser site enable engineers to approximate the distance between the earth and the moon and to estimate the accuracy of this measurement. A milestone in the accuracy of the measurement of the earth–moon distance was the installation, by the Apollo 11 astronauts, of an array of mirrors on the moon's surface. This application will be examined using Maple and the five-step problem-solving process introduced in this chapter.

INTRODUCTION

This chapter begins your introduction to Maple V Release 4, a *computer algebra system (CAS)* developed by the Symbolic Computation Group at the University of Waterloo in Canada. As a CAS, Maple works with mathematical objects such as variables, functions, equations, expressions, sets, lists, and other mathematical quantities, not just the integer or floating-point numbers that are the fundamental data objects used in most spreadsheets, programming languages, and other scientific software packages. This chapter begins with a brief overview of some of Maple's main uses, and then introduces and demonstrates the five-step problem-solving process used in this module. The chapter concludes with some suggestions for maximizing the effectiveness of this module.

1-1 OVERVIEW OF THE MODULE

The broad collection of symbolic, numeric, and graphic tools available in Maple can be used in the mathematical analysis of many engineering applications. Table 1-1 lists examples of common types of manipulations that can be accomplished with Maple. Many of these topics will be discussed in this module. The systematic discussion of Maple as a tool for engineering analysis begins in Chapter 2 with a basic introduction to Maple's powerful graphical user interface, including the online help system. Tools for the definition and symbolic manipulation of expressions and functions are presented in Chapters 3 and 5. This material is separated by a discussion, in Chapter 4, of the graphical capabilities available in Maple. Engineering analysis typically is built upon the concepts of calculus and linear algebra. Maple's capabilities in these areas, as well as its capabilities for the manipulation of objects with complex values, are the primary focus of Chapter 6. The vast majority of Maple is written in the Maple programming language, which is also available to the user. Chapter 7 provides a brief introduction to the Maple programming language for the creation of customized procedures for a variety of problems.

Table 1-1   Overview of Maple Functions

Symbolic Computations
Solution of linear and nonlinear equations and systems
Calculus: limits, derivatives, integrals, series, sums and products
Linear algebra: matrix operations, eigenvalues and eigenvectors, special matrices
Complex numbers, optimization, approximation, statistics, number theory
Piecewise-defined functions, integral transforms, orthogonal polynomials

Numerical Computations
Floating-point computations
Approximate solution of equations and systems (linear and nonlinear)
Numerical integration and solution of differential equations

Graphics and Visualization

Graphs of one or more functions, implicit curves and surfaces, inequalities

Contour, logarithmic, semilogarithmic, log-log, parametric, and polar plots; phase portraits

Animation with playback, looping, direction, and speed controls

---

Maple Programming Language

User viewable and extendible source code for Maple commands

Procedural programming language, similar to standard programming languages

Full-featured debugger with watch points, error trapping, break points, and stepping

---

User Interface

Standard word processing features, including graphics and equations

Collapsible sections and subsections for outlining

Hyperlinks between Maple worksheets

The following example provides a quick illustration of the symbolic, numeric, and graphic operations that can be accomplished with Maple. Commands used in this example will be introduced at the appropriate time later in this module.

EXAMPLE 1-1

The Area of a Circle

The area between two curves can be found by integrating the difference between the two curves over an appropriate interval. For example, the top and bottom of a circle with radius 2 centered at the origin are given by

$y_{top} = \sqrt{2^2 - x^2}$ and $y_{bot} = -\sqrt{2^2 - x^2}$ for $-2 \leq x \leq 2$. Thus, the area of this circle is given by the integral

$$A = \int_{-2}^{2} (y_{top} - y_{bot})\,dx = \int_{-2}^{2} 2\sqrt{2^2 - x^2}\;dx.$$

Use Maple to determine exact and approximate values for this integral and compare the results with the usual formula for the area of a circle.

SOLUTION

A Maple-based approach to this problem begins by assigning names to the expressions that define the top and bottom of the circle:

```
> restart;
> y[top] := sqrt( 2^2 - x^2 );
```

$$y_{top} := \sqrt{4 - x^2}$$

```
> y[bot] := -sqrt( 2^2 - x^2 );
```

$$y_{bot} := -\sqrt{4 - x^2}$$

Note that the powers, roots, and differences are all typed using a standard syntax for scientific expressions. To verify that these assignments are correct, the two functions are plotted:

```
> plot( [ y[top], y[bot] ], x = -2 .. 2, title='Circle with radius 2' );
```

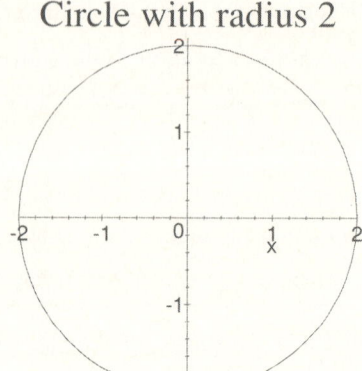

Circle with radius 2

The integral for the area of this circle is being represented as

```
> A := Int( y[top] - y[bot], x = -2 .. 2 );
```

$$A := \int_{-2}^{2} 2\sqrt{4 - x^2}\ dx$$

Note how the previously defined values for **y[top]** and **y[bot]** are substituted into the integral and simplified; the final form is in complete agreement with the integral given. Evaluating this definite integral by manual techniques requires the use of a trigonometric substitution and is somewhat tricky. Maple can compute the value of the integral in a number of different ways. For example, the exact, or symbolic, value is

```
> value( A );
```

$$A := 4\,\pi$$

and the floating-point approximation is

```
> evalf( A );
```

$$12.56637062$$

The symbolic answer is easily seen to be the exact area of a circle with radius 2. It is emphasized that the numerical answer, which is all that can be expected from traditional scientific software packages, is only an approximation.

To summarize, observe the following fundamental characteristics of Maple's syntax:

Maple is case sensitive; for example, the irrational constant π is denoted by **Pi**—the names **pi** and **PI** have no special meaning in Maple

Maple commands end with either a semicolon (**;**) or colon (**:**)

Assignments are designated by **:=** and equations are formed using **=**

At the start of each problem, Maple is reset with the **restart** command.

The important point of this example is the natural way in which Maple can be used to perform symbolic and numeric computations, and produce graphs of mathematical quantities.

This module presents the information essential for the use of Maple. With this background, you should be able to acquire additional skills and techniques via the online help (see Chapter 2) or other printed or electronic resources. The remainder of this chapter presents and demonstrates, by way of examples, the five-step problem-solving process used throughout this module.

1-2   THE FIVE-STEP PROBLEM-SOLVING PROCESS

Engineers are practical problem solvers who use fundamental scientific principles to explain and predict behavior in the real world. In fact, problem solving is the foundation of all engineering activity. In approaching any problem, engineers should always keep the following points in mind:

Question the statement of the problem: Is the problem well defined? What are the important questions? Is all necessary information provided? If not, is it possible to obtain the missing information from other sources?

Gather information related to the problem. Make note of additional information that does not appear to be needed at the outset. Subsequent revisions might require different information.

Think about alternative approaches to solving the problem. Usually more than one approach can be used to solve a problem. Look for special cases where the solution is easy to find. Use a numerical or graphical method to cross-check an exact answer.

Check the units in your answers. Perform an order-of-magnitude analysis of the problem. Then apply successive refinements to obtain solutions that are increasingly more accurate.

Exercise caution since an intuitive answer may not fulfill all the conditions of the current problem.

Plot solutions as well as relevant intermediate results.

Consult team members and other experts to get their perspective on a problem.

Build confidence in your results and your increasing ability to solve problems by explaining and justifying your solutions to others.

This module will help you develop good problem-solving procedures using Maple so you can efficiently accomplish your engineering goals. Table 1-2 summarizes the *five-step problem-solving process* used in this module.

Table 1-2 Maple-Based Five-Step Problem-Solving Process

General Problem-Solving Procedure	Maple Problem-Solving Procedure
1. Define the problem.	Identify questions to be answered.
2. Gather information.	Begin to collect parameters, equations, and other information in a Maple worksheet.
3. Generate and evaluate potential solutions.	Use Maple to assist with the analysis of the problem, including the generation of potential solutions.
4. Refine and implement a solution.	Refine the worksheet. Display results in the form of tables, plots, and so on. Decide whether changes need to be made to data-collection procedures or to Maple analyses.
5. Verify and test the solution.	Verify the solution by performing one or more thorough comparisons using different procedures.

The applications contained in this module employ this five-step process to analyze problems in several different engineering disciplines, as shown in Table 1-3.

Table 1-3 Overview of Applications in This Module

Discipline	Application Problem	Chapter
Engineering	Measuring distances with lasers	1
Mechanical engineering	Strength and toughness of materials	2
Aerospace engineering	Lift and drag	3
Optical engineering	Bandwidth of an optical filter	4
Environmental engineering	Water quality	5
Electrical engineering	Nodal analysis of a circuit	6*
Chemical engineering	Solvents and solutes	7*

* see ftp site

1-3 APPLYING THE FIVE-STEP PROBLEM-SOLVING PROCESS

To understand the five-step problem-solving process, and its use with Maple, study Application 1, which considers the problem of measuring the distance between the earth and the moon. Some of the work is completed with the help of Maple. Although most of these commands are fairly self-explanatory, later chapters will provide explicit instruction in the use of Maple. The main goal for you at this point is to become familiar with the five-step problem-solving process.

Application 1   ## USING PULSED LASERS TO MEASURE DISTANCES

To see how the five-step problem-solving process can be applied, consider the problem of determining the accuracy of a calculated distance between the earth and the moon. This problem has been greatly simplified to allow greater emphasis on the problem-solving procedure.

Fundamentals

The radii of the earth and moon are 6,378 km and 1,738 km, respectively. The first estimates of the distance between the earth and its moon are believed to have been made in the third-century B.C. It was not, however, until 1965 that the estimates were sufficiently accurate, within 600 feet, to be of scientific value. A group of Russian astronomers used a 104-inch telescope to both transmit a 50 ns (50×10^{-9} seconds) ruby laser pulse toward the moon and detect the reflected signal. The surface-to-surface earth–moon distance was estimated by measuring the time required for the signal to complete the roundtrip. The internationally accepted value for the speed of light is c = 299,792.5 kilometers per second, or approximately 1 foot per nanosecond. The curvature and irregularities in the moon's surface impose the greatest limitations on the accuracy of this estimate.

The accuracy of estimates of the earth–moon distance significantly improved when the astronauts of Apollo 11 placed an array of 100 corner reflectors (prismlike devices that reflect a beam of light so that it returns along a path that is exactly parallel to the incident beam) on the surface of the moon. The reflections from these "retroreflectors" have an intensity that is between 10 and 100 times greater than that produced by a reflection from the natural moon surface. A short-pulse laser with durations of 10 ns is directed at the retroreflector array. Even when focused through a 120-inch telescope, the signal has a diameter of 1 mile when it reaches the moon; the reflected signal spreads to almost 10 miles during the return trip to earth. As a result of this diffraction, the receiver detects approximately 25 photons from the 10^{20} photons in the original signal. The roundtrip takes approximately 2.5 seconds, and the estimates based on these measurements are accurate to 6 inches.

The actual earth–moon distance is not as important as the variation in distance as measured over the period of several months or years. Variations in the earth–moon distance are used to determine the distribution of the moon's mass, changes in the location of the earth's poles, and whether the gravitational constant actually is a universal constant.

 ## 1. Define the problem

The purpose of this problem is to determine the relationship between the accuracy of the timing measurement and the accuracy of the distance estimate. Use the information provided to estimate the distance between the earth and moon. How accurate is this estimate? What timing resolution is needed to conclude that the distance is accurate to 6 inches?

◢ 2. Gather information

Some of the background information prior to Step 1 is not directly relevant to the questions of immediate interest. For example, the photon counts and diffraction must be considered when selecting the laser, transmitter, and receiver, but do not influence the measurement of the elapsed time or computation of distance. Note also that units are provided for all quantities.

The speed of light is c = 299,792.5 kilometers per second. If d denotes the surface-to-surface distance in kilometers between the earth and moon, then the center-to-center distance is $d + 8,116$ km. If the elapsed round-trip time is t (approximately 2.5 seconds), the total distance travelled must be the product of the speed and elapsed time: $2d = ct$.

◢ 3. Generate and evaluate potential solutions

Now that the general problem is defined, and most information has been collected, it is time to begin to assemble the information in a form suitable for use with Maple. For this application, you will use Maple primarily as a calculator. You will be asked to convert this information to a Maple worksheet in Problem 10 in Chapter 2.

The earth–moon distance can be found using $d = \dfrac{ct}{2}$.

```
> restart;
> d := c*t/2;
```

$$d := \frac{1}{2}\, c\, t$$

Using the values given for the speed of light and elapsed time

```
> c := 299792.5;
```

$$c := 299792.5$$

```
> t := 2.5;
```

$$t := 2.5$$

the earth–moon distance in kilometers is found to be

```
> d;
```

$$374740.6250$$

There are 0.6214 miles per kilometer. Thus, the earth–moon distance in miles is

```
> d * 0.6214;
```

$$232863.8244$$

4. Refine and implement a solution

Observe that the computational results in Step 3 contain more significant digits than either of the input data. Each computation is only as accurate as the least accurate quantity involved in the computation. Here the quantities that are least accurate are the time, which is known to two significant digits, and the computed distance, which is accurate to only two significant digits. Thus, the surface-to-surface distance found in Step 3 is 370,000 kilometers, or 230,000 miles. Note that with only two significant digits, the distinction between center-to-center and surface-to-surface distance is almost inconsequential.

 ## 5. Verify and test the solution

The last part of the problem is to determine the timing accuracy necessary to estimate the earth–moon distance (d) with a prescribed distance accuracy Δd. Your intuition should suggest that the improved accuracy in the time measurements will lead to improved accuracy in the distance calculations. To determine the precise relationship, note that the elapsed time corresponding to distance d is $t = \dfrac{2\,d}{c}$. Thus, if the actual distance differs by an amount Δd, then the measured elapsed time will be

$$\frac{2\,(d+\Delta d)}{c} \;=\; \frac{2\,d}{c} + \frac{2\,\Delta d}{c} \;=\; t + \frac{2\,\Delta d}{c} \;=\; t + \Delta t$$

where $\Delta t = \dfrac{2\,\Delta d}{c}$ is the corresponding inaccuracy in the elapsed time. Observe that the relationship between Δt and Δd is consistent with our intuition.

Converting 6 inches to kilometers can be accomplished by using the exact conversions of 2.54 cm/inch and 1 cm = 10^{-5} km:

```
> Dd := 6 * 2.54 * 10^(-5) ;
```

$$Dd := .0001524000000$$

which when normalized with a single significant digit is

```
> Dd := evalf( Dd, 1 );
```

$$Dd := .0002$$

The corresponding timing inaccuracy is, to one significant digit:

```
> Dt := evalf( 2*Dd/c, 1 );
```

$$Dt := .1\ 10^{-8}$$

This time is seen to be 1 nanosecond. Observe that this answer is consistent with the remark that light travels approximately 1 foot every nanosecond.

What If

?

Current technology and models can provide regular updates on the distance between the earth and moon. Moon Viewer is a World Wide Web (WWW) tool that provides a user-friendly interface to a variety of information about the moon. The Internet address, or URL, for Moon Viewer is http://saatel.it/users/lore/moon.html. Use a WWW browser such as Netscape or Internet Explorer to access the Moon Viewer homepage. On the web page, fill in today's date, using the format dd-mm-yyyy, click the button labeled "See the Moon," and then record the moon's distance. For example, the moon's distance on July 4, 1997, is 391,284 kilometers. (Your results may vary slightly since these results also depend on the time of day.) How accurate must the timing be to measure this distance to the nearest kilometer? The laser beam would take approximately 2.61037 seconds to complete a roundtrip. Assuming all digits in the moon's distance are accurate, all six digits in the elapsed time are significant. The distance on the day you select may be considerably different. To estimate the average earth–moon distance, find the distance on the first of each month of the year 1997. The smallest distance, 363,688 km, occurs on May 1, and the largest distance, 405,174 km, occurs on October 1. The average distance for the 12 months is (to 6 significant digits) 386,364 km. How do the smallest, largest, and mean distances compare with the (two-significant-digit) distance found in Step 4 of Application 1? Moon Viewer is not specific about whether the distances reported are surface-to-surface or center-to-center. Can this be determined by comparing the distances reported by Moon Viewer with the distances computed from the data in Application 1?

1-4 HOW TO USE THIS MODULE

Learning to use Maple is not a spectator sport. This module should be studied with immediate access to a computer on which Maple is running. You should reproduce all steps described in the text—practice opening and manipulating worksheets, accessing the online help, entering and executing Maple commands, and manipulating Maple graphics. In addition, you are strongly encouraged to implement the solutions to the examples. The Try It! exercises, for which solutions are not provided in the text, are recommended as a first test of your mastery of each topic. Do not be afraid to experiment with Maple and use it to pursue alternative solutions to the ones given in the text and to analyze problems of personal interest.

This module contains explicit instruction on the use of Maple V Release 4. It assumes that you are reasonably familiar with the use of applications running under Windows 95 or NT, including how to use the mouse to point, click, and double-click; start an application from the desktop; navigate the local file structure; perform window operations; and make a menu selection. The appearance and functionality of the worksheet interface should be essentially the same in Windows, Macintosh and UNIX environments.

A few of the examples, Try It! exercises, and end-of-chapter problems refer to auxiliary files that can be downloaded from the Addison-Wesley Toolkit Web site on the WWW. The URL for the Toolkit homepage is http://www.aw.com/cseng/toolkit/.

SUMMARY

This brief chapter introduced a few of the symbolic, numeric, and graphic capabilities of Maple, and presented general guidelines for the use of the five-step problem-solving process with Maple. The general procedure, customized for Maple, can be summarized as follows:

Express the problem in a mathematical form.

Identify variables, collect parameter values, and determine governing equations.

Use Maple to develop a better understanding of the problem and answer questions related to the problem.

Create a Maple worksheet that expresses all relevant information and results in an efficient form.

Cross-check your solution, and apply the solution to answer-related questions.

This process was demonstrated using an application from general engineering that used pulsed lasers to estimate the distance between the earth and moon. The results were cross-checked using data obtained from the Internet.

This chapter concluded with some advice on how to use this module—in particular, that hands-on experience with Maple is critical.

Key Words

computer algebra system (CAS)   five-step problem-solving process

Maple Commands

assignment (:=)   plot
evalf   restart
Int   value

Reference

1. Faller, J.E. and Wampler, E.J. "The Lunar Laser Reflector," *Scientific American,* March 1970, pp. 37–49.

2

Maple Fundamentals

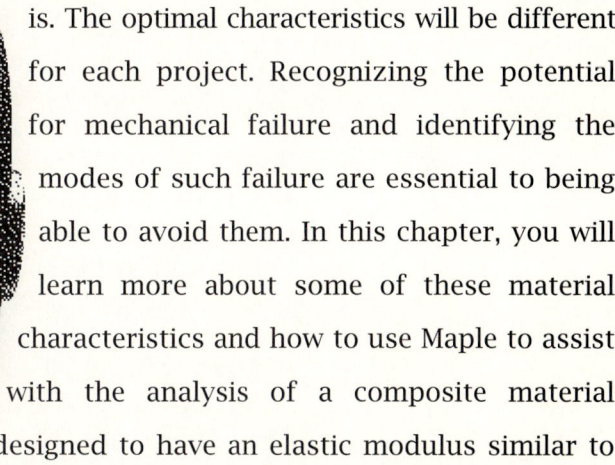

Strength and Toughness of Materials

A mechanical engineer uses a number of criteria to select the materials to be used in a project. These include how much weight the material can support, the rigidity of the complete structure, how much the material properties change with exposure to temperature changes or different types of light, as well as how strong, tough, brittle, or ductile the material is. The optimal characteristics will be different for each project. Recognizing the potential for mechanical failure and identifying the modes of such failure are essential to being able to avoid them. In this chapter, you will learn more about some of these material characteristics and how to use Maple to assist with the analysis of a composite material designed to have an elastic modulus similar to lead and an ultimate strength similar to steel. The modulus of toughness is defined as the strain energy per unit volume stored immediately prior to fracture. The important concepts of stress, strain, and toughness will be defined and discussed using the five-step problem-solving process. Maple worksheets will be used to organize and present your results.

INTRODUCTION

In this chapter, you learn how to use the Maple V Release 4 graphical user interface (GUI), including how to invoke a Maple session, how to create, modify, save, and execute a Maple worksheet, and how to access and navigate Maple's online help system. These concepts are applied in a mechanical engineering application that uses approximations to the area under a curve to estimate the toughness of a composite material. The chapter concludes with a presentation of some of the features you can use to create more efficient and visually appealing worksheets. The systematic introduction to Maple commands that begins in Chapter 3 uses the techniques and terminology established in this chapter.

You should read this material in a setting where you have immediate access to a computer and Maple. You are strongly encouraged to follow the step-by-step directions as they are presented. Illustrations showing the current state of your worksheet are included to help you check your progress. The numerous Try It! and other exercises should be solved as you encounter them. At the completion of this chapter, you will have the fundamental skills necessary to acquire the material presented in the other chapters and to apply these techniques in subsequent mathematics and engineering courses.

If you have access to an earlier version of Maple V, you should be able to follow most of the discussion in this module. Even though significant changes have been made to the worksheet interface, the syntax of most Maple commands (particularly those encountered in this module) has not changed.

2-1 WORKING WITH MAPLE WORKSHEETS

One of the biggest improvements in Maple V Release 4 is the complete redesign of the GUI. The *graphical user interface* closely resembles the interface for many Windows-based word processors, spreadsheets, and other applications. Many of the standard operations, such as New, Open, Print, Cut, Copy, and Paste, are accessed via the same menus, icons, and keyboard shortcuts that are used in many of the leading Windows applications. This module assumes that you are familiar with these common operations. Additional information about the Maple interface can be found in Chapter 1 of the *Maple V Learning Guide* or by using the online help to conduct a topic search for the word *worksheet*. (Additional techniques for using the online help are discussed in Section 2-2.)

The best way to learn about *Maple worksheets* is to jump in and experiment with Maple. Since the purpose of this chapter is to provide an introduction to some of the fundamental techniques for working with Maple worksheets, we will not fully explore each of the commands that are used. This should present no problems, however, because the command names describe the operations that they perform. For example, the **solve** command returns the solution(s) to a system of equations; **subs** substitutes a value for a parameter in an expression; **simplify** attempts to simplify an expression; **plot** creates a plot of a function on a given interval; and **eval** and **evalf** are used to force the evaluation of an expression. A full description of these and many other Maple commands is contained in the subsequent chapters of this module.

In this section, you will learn about the different parts of the Maple window and the techniques necessary to create a Maple worksheet containing input commands, text, and graphics. You should perform each step as it is presented. Feel free to explore different options as your interest increases. At several different points in the discussion, you can compare your worksheet with samples that are provided with this text.

Since the emphasis of this chapter is on the worksheet fundamentals, neither the mathematics nor the engineering will be too sophisticated. You will practice using the Maple worksheets by solving the following *tangency problem*: For what values of the parameter a do the two graphs of $f(x) = x^3 + x$ and $g(x) = ax^2$ intersect at a point where their tangent lines coincide? The slope of the tangent line to a curve is given by the derivative: $f'(x) = 2x^2 + 1$ and $g'(x) = 2ax$. (If you have not yet had calculus, ask a friend what a derivative is. Calculus is not used in an essential way in this module until Chapter 6.) Thus, to solve this problem, it is necessary to determine all values of the parameter a for which the equations $f(x) = g(x)$ and $f'(x) = g'(x)$ are satisfied for the same value of x. Primed quantities represent the first derivative.

Getting Started

Almost all interaction with Maple can be performed through the Maple GUI and Maple worksheets. The first step in preparing to use Maple is to launch the application by selecting Maple V Release 4 from the Programs menu. This starts a Maple session with a blank Maple worksheet.

You can also start a Maple session by double-clicking the icon or filename for a Maple worksheet in a folder of files. A Release 4 worksheet can be identified by a `.mws` extension; earlier versions of Maple use `.ms`. In this case, the Maple session starts as before, but displays the contents of the selected worksheet.

Try It Start a Maple session with an empty worksheet. The name of the worksheet should be *Untitled* (1). Use one of the standard methods for your platform to maximize the worksheet (that is, expand the Maple worksheet so that it completely fills the window). The result should be very similar to Figure 2-1.

Figure 2-1
An empty Maple worksheet

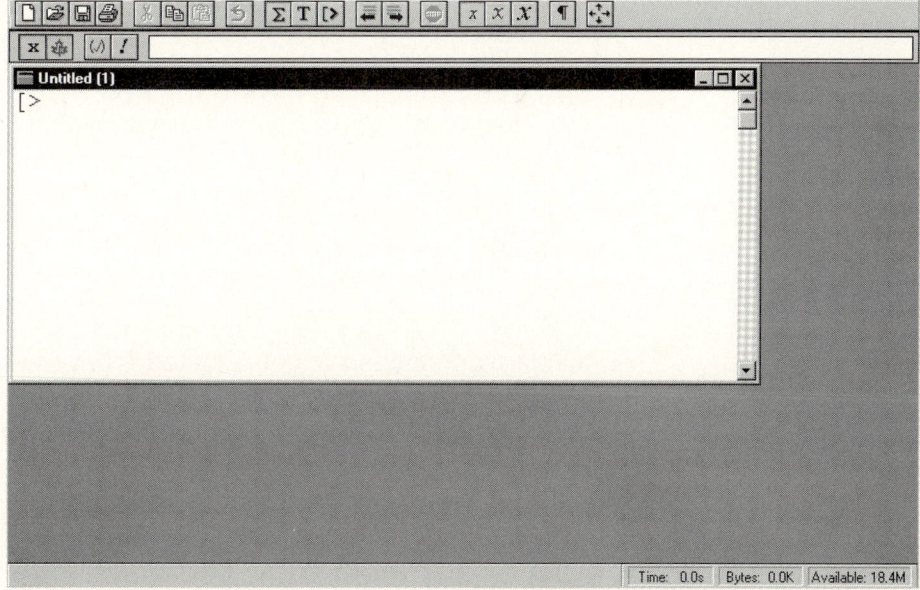

Anatomy of a Worksheet

Figure 2-2 shows the main features of the Maple interface. The *menu bar* is immediately below the *title bar* at the top of the Maple window. Although the collection of menus available from the menu bar changes depending on the context, as determined by the current location of the cursor, the File, Edit, View, and Help menus are almost always available. A written description of the currently highlighted menu selection is displayed in the *status line* at the bottom of the Maple window. The status line also contains information about memory and CPU usage.

Figure 2-2
The Maple interface

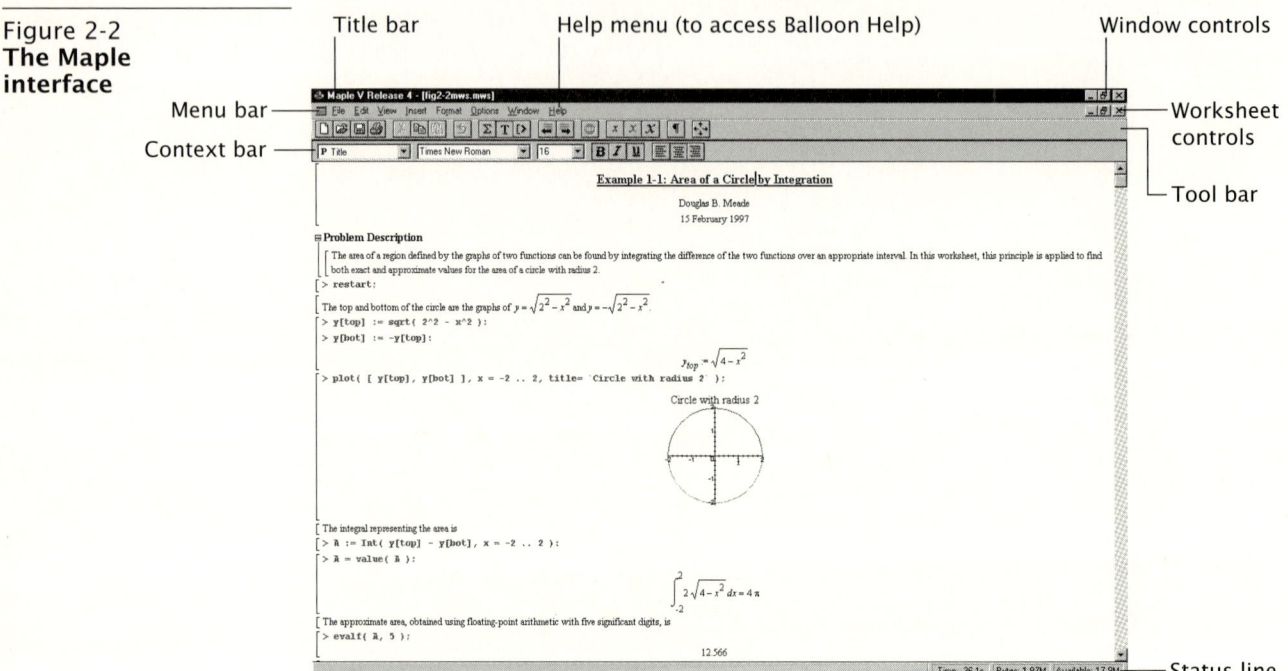

Title bar · Help menu (to access Balloon Help) · Window controls · Menu bar · Context bar · Worksheet controls · Tool bar · Status line

The *tool bar,* located immediately below the menu bar, contains a number of icons that provide immediate access to some of the more commonly used entries from the menus in the menu bar.

Balloon help is a particularly useful feature when you are learning the operations associated with the icons. To activate this feature, select Balloon Help from the Help menu on the menu bar. Then, whenever the cursor is located on an icon, a brief description of the operation performed by this icon is displayed next to the icon.

Try It   Activate balloon help in your current session. Then, use it to display the descriptions of each of the icons in the toolbar.

Maple worksheets are constructed from five different types of regions: text, input, 2D graphics, 3D graphics, and animation. (Note that output regions are actually text regions displayed in a predefined style.) All Maple commands are entered in *input regions,* which can be identified by the **>** at the left edge of the worksheet. Specific information for the manipulation of different types of regions will be explained as needed later in this chapter and throughout the remainder of the module. The specific collection of buttons and information displayed in the *context bar,* located directly below the tool bar, depends on the current context, as defined by the type of region in which the cursor is located. Regions and some of the ways they can be combined are discussed in Section 2-3.

The remainder of the Maple window is reserved for displaying one or more Maple worksheets. Each worksheet can be iconified, resized, or closed using the buttons at the extreme right end of the menu bar. These buttons should not be confused with the similar-looking buttons on the Maple window. The buttons on the worksheet pertain only to the corresponding worksheet; the buttons on the Maple window control the entire Maple session.

The user can control the display of the tool bar, context bar, and status line from the first three selections in the View menu. One reason for hiding one or more of these bars is to increase the screen area available for displaying a Maple worksheet during a presentation.

Try It Use the toggle switches at the top of the View menu to see how the interface changes when various menu bars are hidden from view. When finished, be sure that all bars are visible.

Worksheet Fundamentals

Begin with an empty worksheet, as illustrated in Figure 2-1. If your window does not look like this, you can display a new worksheet by clicking the leftmost icon in the tool bar (or by selecting New from the File menu). You are strongly encouraged to begin all worksheets by resetting Maple. This is done by entering the **restart** command at the input prompt. To execute the command, press the Enter key. Recall from Chapter 1 that the end of a command is marked by a semicolon (;) or colon (:).

```
> restart;
```

The first step is to define the two functions mentioned in the statement of the problem. *Assignments* are made using the composite operator :=.

```
> f := x^3 + x;
```

$$f := x^3 + x$$

```
> g := a*x^2;
```

$$g := a\,x^2$$

Note that only the function name is used on the left-hand side of the assignment, f, not f(x). The points of intersection of these functions can be found by solving the equation f(x) = g(x) for the variable *x*:

```
> SOLa := [ solve( f = g, { x } ) ];
```

$$SOLa := \left[\{x = 0\}, \{x = \frac{1}{2}\,a + \frac{1}{2}\sqrt{a^2 - 4}\}, \{x = \frac{1}{2}\,a - \frac{1}{2}\sqrt{a^2 - 4}\} \right]$$

Observe the different uses of `:=` and `=` in the previous command. There are three solutions to this equation because all cubic polynomials have three roots (counting multiplicities of repeated roots). Two of the three roots depend on the parameter a. To see the roots for specific choices of the parameter, the **subs** command can be used. Here, the roots corresponding to $a = 2$ and $a = 3$ are obtained:

> `SOLa2 := subs( a = 2, SOLa );`

$$SOLa2 := [\{x = 0\}, \{x = 1\}, \{x = 1\}]$$

> `SOLa3 := subs( a = 3, SOLa );`

$$SOLa3 := \left[\{x = 0\}, \{x = \frac{3}{2} + \frac{1}{2}\sqrt{5}\}, \{x = \frac{3}{2} - \frac{1}{2}\sqrt{5}\} \right]$$

These are the exact values of the roots. The **evalf** command can be used to obtain the floating-point approximation of the roots to any number of significant digits. If the number of digits is not explicitly specified, Maple uses 10 digits by default:

> `evalf( SOLa3 );`

$$[\{x = 0\}, \{x = 2.618033989\}, \{x = .381966011\}]$$

Two of Maple's basic data types are lists and sets. A *list* is an ordered collection of Maple objects (names, numbers, equations, and so on) delimited by square brackets (`[ ]`). A *set* is the same, except the elements are not ordered, and the objects are delimited by curly braces (`{ }`). For example, each solution to the equation here is returned as a set, and a list is formed from each collection of three solution sets. This particular choice is made so that, for example, you can refer to the "second solution to $f = g$" as `SOLa[2]` and not be concerned with the order in which Maple might present the elements in a set. Additional examples where the distinction between ordered and unordered collections of objects are important will be discussed in more detail in Chapter 3.

The **diff** command computes the derivative of a function with respect to the specified variable. Thus,

> `df := diff( f, x );`

$$df := 3\,x^2 + 1$$

> `dg := diff( g, x );`

$$dg := 2\,a\,x$$

The same process can be used to find the intersections of the derivatives of the functions. Since $f(x)' = g(x)'$ is a quadratic equation, there should be two points where the functions have the same slope. In fact,

```
> SOLb := [ solve( df = dg, { x } ) ];
```

$$SOLb := \left[\{x = \frac{1}{3}\, a + \frac{1}{3}\sqrt{a^2 - 3}\, \}, \{x = \frac{1}{3}\, a - \frac{1}{3}\sqrt{a^2 - 3}\, \} \right]$$

When $a = 0$, the roots are

```
> subs( a = 0, SOLb );
```

$$\left[\{x = \frac{1}{3}\sqrt{-3}\}, \{x = -\frac{1}{3}\sqrt{-3}\} \right]$$

Note the square root of a negative number. Although complex-valued solutions are not appropriate for this problem, it is good to realize that Maple is able to deal with such a problem in a natural way. They arise for $a = 0$ because the tangent for f(x) never coincides with the abscissa for any x. The floating-point approximations obtained by **evalf** are:

```
> evalf( " );
```

$$[\{x = .5773502693 \, I\}, \{x = -.5773502693 \, I\}]$$

The double quotation marks (") refer to the previous result; in the same way "" and """ refer, respectively, to the second and third most recent results. Observe that though mathematicians and physicists commonly use i to denote the imaginary unit, that is, $i = \sqrt{-1}$, engineers typically use j and Maple uses **I**.

To conclude the problem, it is necessary to determine if any values of a give the same root for the two different equations. It might be possible to find a solution to this problem by substituting different values of a into both **SOLa** and **SOLb**, but this is not likely to work for most problems. A better approach is to simultaneously solve the two equations **f = g** and **df = dg** for both x and a.

```
> SOL := solve( { f = g, df = dg }, { x, a } );
```

$$SOL := \{a = 2, x = 1\}, \{x = -1, a = -2\}$$

This shows that Maple has found two solutions to the problem. It is a good habit to check that Maple's results actually solve the desired problem. Here, this means verifying that the equations are satisfied for each of the two solutions.

```
> subs( SOL[1], [ a, x, f = g, df = dg ] );
```

$$[2, 1, 2 = 2, 4 = 4]$$

Note that **SOL[1]** refers to the first of the two solutions and that when $a = 2$ the two functions and their derivatives intersect at the point (1, 2) with slope 4.

The second solution can be verified by replacing the **1** in **SOL[1]** with a **2**. (See Problem 1 at the end of this chapter.) The simplest way to make this change is to use the mouse to select the **1** and type **2**. Then reexecute the command. The drawback to this technique is that the results from the first verification are no longer visible. It is preferable to have both verifications in the worksheet. Fortunately, it is not necessary to retype the command. First, select the **subs** command by either dragging the mouse over the command or triple-clicking anywhere in the command. Use Copy in the Edit menu to put this selection into the clipboard, reposition the cursor in an empty input region, and select Paste from the Edit menu. You can now use the original method to edit the copied command.

Try It Use the Copy and Paste method just described to complete the verification that the second solution satisfies the two equations.

At this point your worksheet should look very similar to Figure 2-3.

Figure 2-3
The worksheet
first.mws.

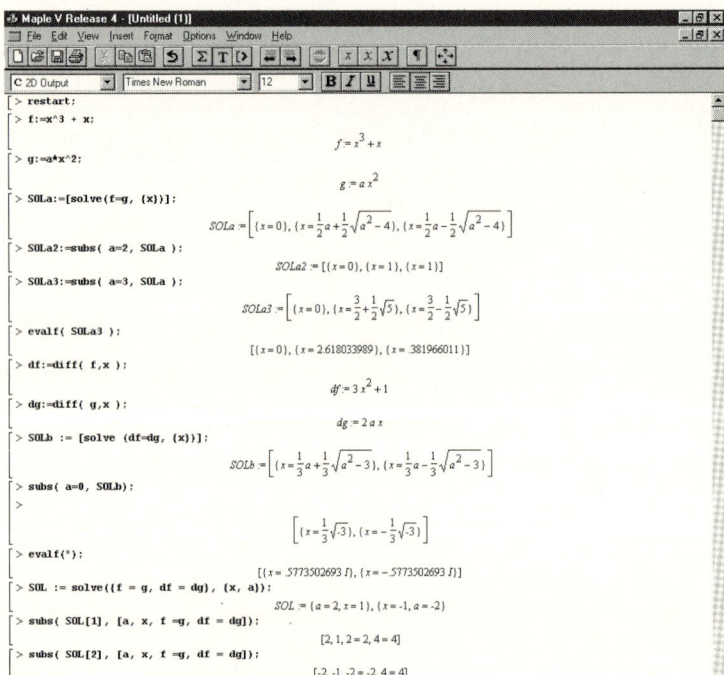

Saving and Loading Worksheets

Maple does not automatically back up your work. If the computer or Maple should ever crash, you will lose all of your work since the most recent save. For this reason, you should make it a habit to save your work often. To save a worksheet, select Save (or, if you wish to change the name of the worksheet, Save As) from the File menu. Since this is a new worksheet, you will be asked to provide a name for the worksheet.

Try It Save the current worksheet with the name *myfirst.mws.*

To load a worksheet that has been saved previously, select Open from the File menu. Highlight the desired worksheet in the dialog box that appears, and then click OK.

Try It   Open the worksheet *first.mws* located on the Toolkit homepage. Compare this worksheet with the one you have just created, *myfirst.mws*.

When a worksheet is opened, the cursor will be located wherever it was when the worksheet was saved. It is a good habit to move the cursor to the top of the worksheet before saving a worksheet.

Note that loading a worksheet does not execute the commands in the worksheet. Even though the worksheet may contain Maple output, these results are not known to Maple until the corresponding commands are executed in the current session.

Managing Multiple Worksheets

A powerful, yet potentially confusing, feature of Maple is that all Maple worksheets opened in one Maple session access a single Maple kernel. This means that any assignment made in one worksheet can be accessed by any other worksheet opened in the same Maple session.

Try It   To explore the ramifications of the fact that all worksheets within the same Maple session access the same Maple kernel, create a new worksheet (the default name should be *Untitled (2)*). Enter and execute the command **f;** in the new worksheet. The result should be the expression representing the function f that was entered in the worksheet now titled *myfirst.mws*. Note that even though there is nothing in *Untitled (2)* to indicate how **f** received a value, the result clearly shows that *Untitled (2)* is sharing information with *myfirst.mws*, and any other command executed within this Maple session.

Very few situations justify the use of multiple worksheets in a single Maple session. For this reason, we suggest that, except when copying commands between worksheets, no more than one worksheet be opened at any one time. Users who wish to open multiple documents, each with its separate Maple kernel, should select the Parallel Server Maple V executable from the Programs menu when launching Maple.

Working with Graphics Regions

Your initial exposure to Maple worksheets has emphasized symbolic and numeric features. To see the basic graphical capabilities available in Maple, you can graphically verify the symbolic results obtained in the worksheet *myfirst.mws*. Be sure that *myfirst.mws* is the only open worksheet and that all the commands have been executed. (A quick way to execute an entire worksheet is to select Execute Worksheet from the Edit menu.) The graphical verification of the solution will be conducted by plotting f, g (for one of the two values of *a*) and the derivatives f′ and g′. Begin by constructing a new function **G2** that corresponds to g when *a* = 2:

```
> G2 := subs( a=2, g );
```

$$G2 := 2\,x^2$$

The derivative can be obtained either by differentiating the new function or by substituting $a = 2$ into the previously computed derivative, g'. In either case

```
> dG2 := subs( a=2, dg );
```

$$dG2 := 4\,x$$

The **plot** command is used to plot all four functions on an interval containing the common intersection at $x = 1$.

```
> plot( [ f, G2, df, dG2 ], x=-2..2,
>         title='Graphical verification when a=2' );
```

Graphical verification when a=2

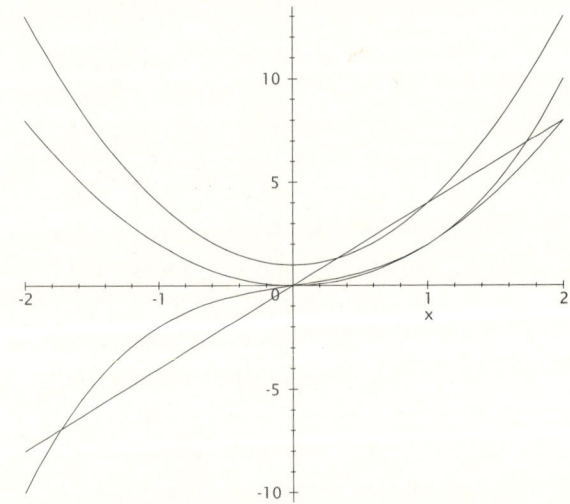

Your plot should use different colors for the different functions. (In Chapter 4 you will learn how you can improve the appearance of this plot.) Observe that there are two pairs of curves that intersect at $x = 1$ and that the lower pair of graphs, the ones corresponding to **f** and **G2**, intersect and are tangential at $x = 1$. This is a graphical interpretation of the equality of the derivatives at the point of intersection. The other intersection of **f** and **G2**, $x = 0$, is obviously not a point of tangency.

The derivatives also have a second point of intersection. The (approximate) coordinates of this point can be obtained directly from the graph. Begin by making the *graphics region* active by clicking the left mouse button when the cursor is located anywhere over the graph. The menu and context bars change to reflect the new current context (see Figure 2-4). When the current context is a graphics region, the leftmost entry in the context bar contains the coordinates of the point where the mouse was last clicked. Thus, to obtain the approximate location of the second point of intersection of **df** and **dG2**, simply move the cursor directly above the point of intersection and click the left mouse button. In this case, we find the point to be approximately (0.3416, 1.201).

Try It

Use the **subs** command to determine the exact location of the second point of intersection for the derivatives. Are the answers you obtained in reasonable agreement symbolically and graphically? Why is this point not a solution to the tangency problem?

Documenting a Worksheet

The approach used to solve a problem is usually self-explanatory to the author at the time it is written. However, if you return to one of your worksheets in the future, or if your worksheet is read by others, the specific steps may not be so clear. All worksheets should include a title, the names of all authors, an overview of the problem to be solved, a description of each step in the solution, and a clear answer to the original question.

Title and author information should appear at the very top of the worksheet. To insert a *text region* at the top of the worksheet (before the **restart** command), place the cursor in the first line of the worksheet, select Execution Group from the Insert menu, and choose Before Cursor. Note that the new region is an input region. To convert the new region to a text region, select Text Input from the Insert menu. In Figure 2-4, you'll notice the changes in the menu and context bars that occur when a region is converted from an input region to a text region and back (by selecting Maple Input from the Insert menu).

Figure 2-4
Comparison of menu, tool, and context bars for (a) input, (b) text, and (c) 2D graphics regions

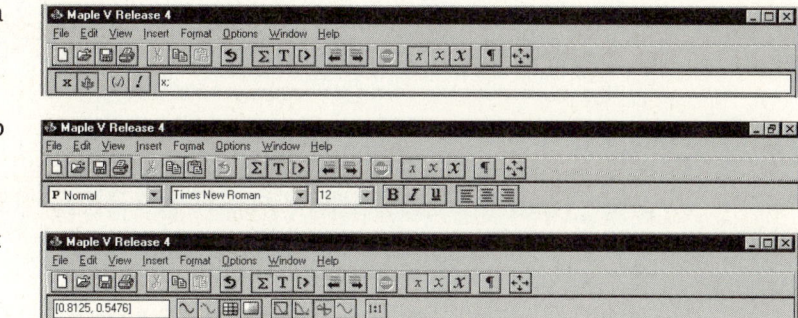

Maple has predefined styles for many common elements of a technical document. The default style is Normal. To indicate that the first line should be the title of the worksheet, select Title from the menu at the left-most end of the context bar. Type the title of this worksheet: for example, **My First Documented Maple Worksheet**. End the title by pressing Enter. Note that the style of the next line has been automatically changed to Author. Enter the names of all who contributed to the preparation of the worksheet. Note that subsequent lines return to the Normal style.

Style changes can also be made after the text has been entered. Simply highlight a block of text, and then select a style from the appropriate menu in the context bar. Changes to the font and font size are made in the same manner. Use these techniques to insert the date on which this worksheet was created on a centered line of italic text directly below the author's name.

You can also add descriptions of the individual steps in the analysis of the problem. To insert after the current region, select Execution Group from the Insert menu, and then choose After Cursor. Change the region from an input region to a text region, and add text as described earlier.

Try It

The insertion of an execution group after the cursor and conversion of a region between text and input are used so often that shortcuts are provided on the tool bar. Use balloon help to locate the icons on the tool bar that correspond to a) inserting a new execution group after the cursor, b) inserting and formatting inert text, and c) inserting Maple commands in a text region.

EXAMPLE 2-1

Documenting a Worksheet

Use the techniques introduced in this section to document the major steps in the solution of the tangency problem. Use Save As to save your documented worksheet with the name *mysecond.mws*.

SOLUTION

The finished worksheet should closely resemble Figure 2-5.

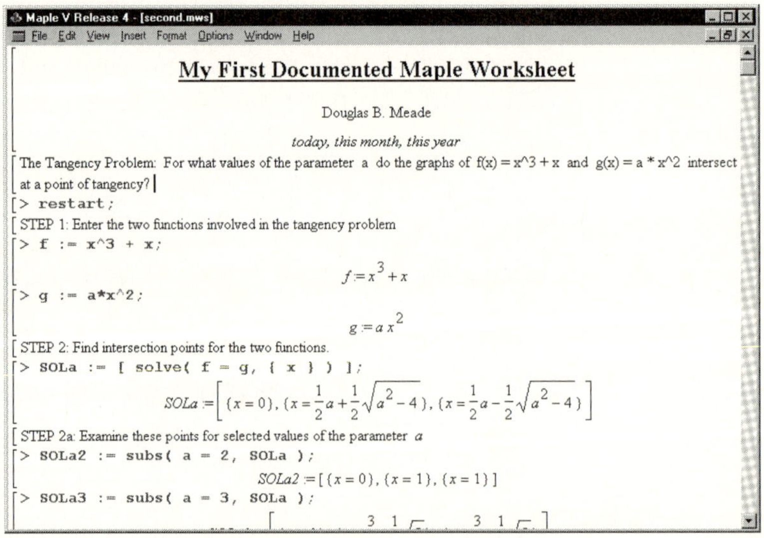

• •

Figure 2-5
Documented worksheet for the tangency problem

Maple V Release 4 - [second.mws]

File Edit View Insert Format Options Window Help

My First Documented Maple Worksheet

Douglas B. Meade

today, this month, this year

The Tangency Problem: For what values of the parameter a do the graphs of f(x) = x^3 + x and g(x) = a * x^2 intersect at a point of tangency?

`> restart;`

STEP 1: Enter the two functions involved in the tangency problem

`> f := x^3 + x;`

$$f = x^3 + x$$

`> g := a*x^2;`

$$g = a\,x^2$$

STEP 2: Find intersection points for the two functions.

`> SOLa := [ solve( f = g, { x } ) ];`

$$SOLa = \left[\{x = 0\}, \left\{x = \frac{1}{2}a + \frac{1}{2}\sqrt{a^2 - 4}\right\}, \left\{x = \frac{1}{2}a - \frac{1}{2}\sqrt{a^2 - 4}\right\} \right]$$

STEP 2a: Examine these points for selected values of the parameter *a*

`> SOLa2 := subs( a = 2, SOLa );`

$$SOLa2 = [\{x = 0\}, \{x = 1\}, \{x = 1\}]$$

`> SOLa3 := subs( a = 3, SOLa );`

2-2 USING ONLINE HELP

The initial examples have demonstrated that Maple commands generally have descriptive and easy-to-remember names. However, even after you become an experienced Maple user, there may be times when you need help remembering the name of a command, its basic syntax, or the optional arguments that produce a result in the desired form. Maple's online *help* provides a complete, and easy-to-access, source of this information. The help information for a specific command can be accessed via the Maple command **help** (or **?**) or the Help menu on the menu bar.

Each help page is actually a separate Maple worksheet. For example, the help information for the word *assignment* is shown in Figure 2-6. The typical help page begins with the name of the command, a brief description of the contents of this help document, and, if appropriate, a summary of the syntax. This is immediately followed by a detailed description of the

Figure 2-6
Online help for Maple's assignment statement

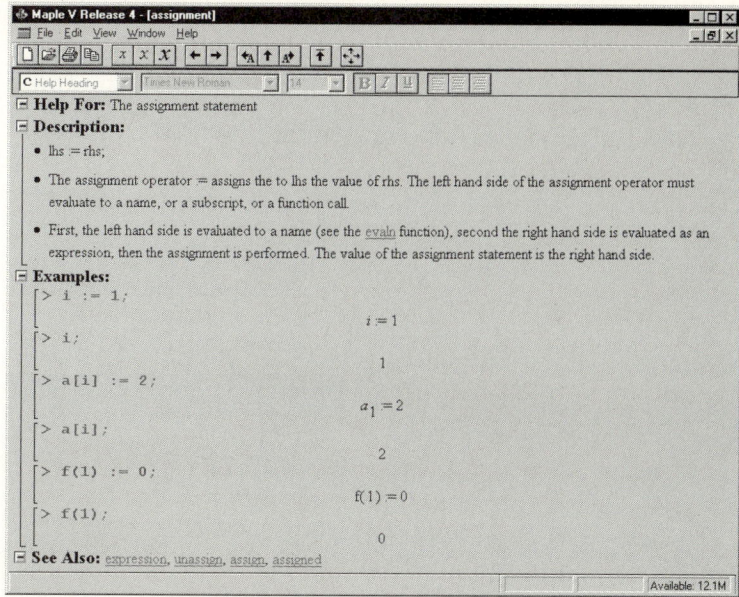

command with particular attention to any input arguments and the result returned by the procedure. The next section in the help page is a collection of examples (with output) that illustrate typical uses of the procedure. Note that these examples cannot be executed or modified in the help worksheet. To experiment with the commands, select Copy Examples from the Edit menu corresponding to the help worksheet, and then paste the examples into an active worksheet. The final section of each help page is a list of cross-references. In most cases, the items in this list are hyperlinked to the appropriate help page—that is, clicking a word that is displayed in green will open the hyperlinked document.

Try It The online help page that describes most features of the Maple worksheet can be accessed using **help(worksheet);**. Load this help document. Click the highlighted string Help System Guide. This opens another help document. Locate and read the information about searching the table of contents.

In addition to searching for help from the table of contents, you can search by topic or by selected words. For instructions on the use of these features, see the help for worksheets displayed in the preceding Try It! exercise.

EXAMPLE 2-2   ## Searching for Help Information

Compare the results of a topic search and a full text search to locate information on Maple worksheets.

SOLUTION

All help topics that begin with *worksheet* are obtained as soon as you type **wo** in the topic box in the Topic Search dialog box (see Figure 2-7). To obtain a list of all help documents that contain the word *worksheet*, type the word **worksheet** in the Word(s) box in the Full Text Search dialog box, and then click the Search button (see Figure 2-8).

Figure 2-7
The Topic Search
dialog box

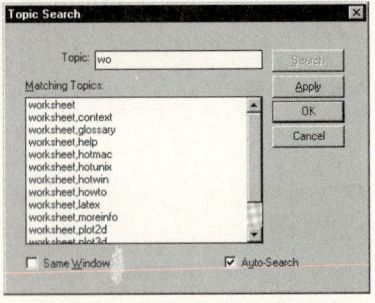

Figure 2-8
The Full Text Search
dialog box

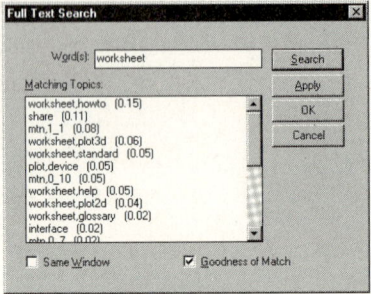

- -

EXAMPLE 2-3

Help for the Student Package

The *student package* is a collection of Maple procedures that provides extra functionality frequently needed by students. Find the general description of the student package. What command must you enter before you can access any of the commands in the student package?

SOLUTION

There are a number of ways to access the information about the student package. A topic search with a search string of **stu** provides a list of more than 30 different help documents that appear to be related to the student package. Of these, the first appears to be the most general. Sure enough, this page lists all procedures contained in the student package. It also informs you that the command **with(student);** must be entered before using the commands from the student package. Note that an even longer list of matches is returned from a full text search for help documents containing the word **student**. Note that the same information is also available by entering the command **help(student);** or **?student**.

You should be aware of one additional method of accessing help documents. The current location of the cursor defines what is known as the *current context*. The current context is listed, in quotes, in the second selection under the Help menu. (If there is no current context, then Help on Context appears dimmed in the menu.) Context-based help can be a very efficient method for learning about commands that appear in worksheets.

- -

Try It

Place the cursor on the word **plot** in the plot command toward the bottom of the *mysecond.mws* worksheet. Use context-based help to access the help for the plot command. Read this information, and the examples, to find out how to modify the **plot** command to cause f and f′ to be displayed in blue and the graphs of g and g′ in red.

Application 2   ## STRENGTH AND TOUGHNESS OF MATERIALS

Mechanical engineers have the essential task of developing new materials that have specific properties. Examples of the characteristics used to describe a material are strength, toughness, corrosion resistance, and fatigue resistance. The selection of the appropriate material is based on its suitability for a particular application, including environmental, economic, and ergonomic (human engineering) factors. This application introduces materials concepts such as elastic (or Young's) modulus, ultimate strength, and fracture strength. In engineering, strength and toughness have very specific, and different, meanings. A strong material will have high ultimate strength, whereas a tough material will withstand a significant stress before breaking. Many of these characteristics can be determined from the stress–strain curve for the material. In this application, the stress–strain curve will be used to estimate the toughness of a composite material.

Fundamentals

If the ends of a bar with length L and cross-sectional area A are pulled in opposite directions with a force F, the bar will stretch. Let ΔL denote the change in the length of the bar. The force per unit area is the stress,

$\sigma = \dfrac{F}{A}$, and the strain is the relative change in length of the material,

$\varepsilon = \dfrac{\Delta L}{L}$.

For small deformations, the stress is linearly proportional to the strain. This is Hooke's law: $\sigma = E\varepsilon$ where the constant of proportionality, E, is the *elastic*, or *Young's, modulus* of the material. Since less force is generally needed to elongate the material at higher temperatures, the Young's modulus typically decreases as temperatures increase. A *hard* material has a high modulus of elasticity, and small values of E indicate a *soft* material. The Young's moduli for some common materials, at room temperature, are given in Table 2-1.

Stress, since it is a force per unit area, is measured in thousands of pounds per square inch, or ksi = kip/in^2, where 1 kip = 10^3 pounds. The strain, which is dimensionless, is usually reported as a percentage. The Young's modulus is generally given in MPa = 10^6 Pascal = 10^6 Newtons/m^2. To convert between units, 1 MPa = 145.14 lb/in^2 = 0.145 ksi.

**Table 2-1   Approximate Young's Modulus
of Representative Materials at Room Temperature**

material	lbf/in^2	MPa
aluminum alloys	$10\text{–}11\times10^6$	$7\text{–}8\times10^4$
brass	$15\text{–}16\times10^6$	$10\text{–}11\times10^4$
cast iron	$15\text{–}22\times10^6$	$10\text{–}15\times10^4$
cast iron, ductile	$22\text{–}25\times10^6$	$15\text{–}17\times10^4$
cast iron, malleable	$26\text{–}27\times10^6$	$18\text{–}19\times10^4$
copper alloys	$17\text{–}18\times10^6$	$11\text{–}12\times10^4$
glass	$7\text{–}12\times10^6$	$5\text{–}8\times10^4$
magnesium alloys	6.5×10^6	4.5×10^4
molybdenum	47×10^6	32×10^4
nickel alloys	$26\text{–}30\times10^6$	$18\text{–}21\times10^4$
steel, hard	30×10^6	21×10^4
steel, soft	29×10^6	20×10^4
steel, stainless	$28\text{–}30\times10^6$	$19\text{–}21\times10^4$
titanium	$15\text{–}17\times10^6$	$10\text{–}11\times10^4$

(Multiply lbf/in^2 by 6.89×10^{-3} to obtain MPa.)

Engineers typically present the stress–strain relationship for a material by plotting stress as a function of strain (see Figure 2-9). Although
this may seem unnatural, note that different strains can produce the
same stress in the material. Thus, mathematically speaking, strain is
not always a single-valued function of stress.

Figure 2-9
**Typical stress–strain
curve for a material**

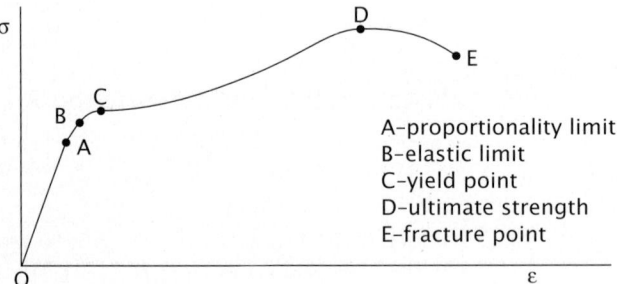

A–proportionality limit
B–elastic limit
C–yield point
D–ultimate strength
E–fracture point

The *ultimate strength* of a material (point D in Figure 2-9) is the maximum stress the material can support. The *fracture point* of a material
(point E) is the stress and strain at which the material actually fails.
A brittle material is characterized by a fracture point at low strain, typically within the linear portion of the stress-strain curve; ductile materials
can withstand a large stress prior to failure.

A *tough* material can withstand occasional high stresses without
fracturing. The strain energy per unit volume stored in a specimen
when it reaches incipient fracture, or *modulus of toughness* of a material, is
one indication of a material's toughness. The modulus of toughness is
given by the total area under the stress–strain curve. The area under a
curve can be expressed as an integral. However, since the stress–strain

relationship is typically obtained from experimental data, no explicit function may express stress as a function of strain. Thus, the area under the curve often needs to be approximated.

The five-step problem-solving process will be used to calculate the modulus of toughness of a material whose stress–strain curve is displayed in Figure 2-10. You should implement all steps of this solution, including the written description of the problem. You are also encouraged to collect your own data points if you do not agree with the estimates used in the solution. What impact does this have on the final analysis?

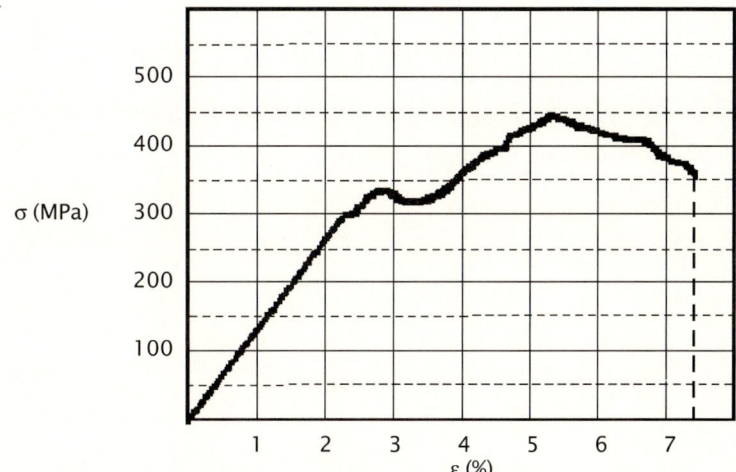

Figure 2-10
Stress–strain measurements for a composite material

 1. Define the problem

The primary objective is to calculate, approximately, the modulus of toughness for the material whose stress–strain curve is provided. In completing this task, the elastic modulus, ultimate strength, and fracture point should all be determined.

 2. Gather information

An experimentally measured stress–strain curve for a composite ductile material is given in Figure 2-10. The material has been designed by an engineer to have an elastic modulus similar to lead, but an ultimate strength similar to steel. This single plot contains all the information needed to analyze the material. Note that since the data is obtained from an experiment, it contains an unknown amount of measurement error. One indication of the experimental nature of this data is the oscillation in the stress–strain curve. The thickness of the recorder pen also contributes to inaccuracies in obtaining data from Figure 2-10. This uncertainty will have to be incorporated into the analysis in terms of the number of significant figures used in subsequent calculations.

 3. Generate and evaluate potential solutions

You will determine initial approximations for each of the requested quantities. More accurate approximations will be explored, as needed, in the final steps of the process. The stress–strain curve can be divided into three disjoint sections: (1) the elastic region with $\varepsilon \le 2.8\%$ where Hooke's law applies; (2) the relatively constant region, $2.8\% \le \varepsilon \le 3.2\%$, where stress is relatively constant with increasing strain (typical of a soft and weak material); and (3) the concave-down region that begins near $\varepsilon = 3.2\%$ and ends with failure at $\varepsilon = 7.4\%$.

The elastic modulus of this material is obtained by estimating the slope of the elastic portion of the stress–strain curve. To estimate this slope, determine two points on this part of the curve, and then use Maple to compute the slope between these two points.

Suppose the two points at the endpoints of the elastic region are estimated to be (0, 0) and (2.8%, 330 MPa) or, in Maple,

```
> restart;
> pt1 := [0,0];
```

$$pt1 := [0, 0]$$

```
> pt2 := [0.028,330];
```

$$pt2 := [.028, 330]$$

The slope of the line between these two points is computed to be

```
> E := ( pt2[2]-pt1[2] ) / ( pt2[1] - pt1[1] );
```

$$E := 11785.71429$$

Truncating this computation to two significant digits, the Young's modulus is approximately $E = 1.2 \times 10^4$ MPa, similar to lead.

The next segment of the graph, for strains between 2.8% and 3.2%, is nonlinear. However, because this is a relatively narrow band of strains and because the stress values do not vary too much, a constant function can be used to approximate this portion of the curve without significantly affecting the estimation of the modulus of toughness. The endpoints of this region are (2.8%, 330 MPa) and (3.2%, 330 MPa). The new data point is

```
> pt3 := [ 0.032, 330 ];
```

$$pt3 := [.032, 330]$$

For this material, the ultimate strength is seen to be approximately 440 MPa, which occurs when the strain is approximately 5.3%:

```
> pt4 := [ 0.053, 440 ];
```

$$pt4 := [.053, 440]$$

The fracture strength is the stress that is observed immediately before failure. For this material, failure occurs near $\varepsilon = 7.4\%$. The corresponding stress (the fracture strength) is 360 MPa. The fracture point is, therefore,

```
> pt5 := [ 0.074, 360 ];
```

$$pt5 := [.074, 360]$$

These observations provide two-digit estimates of all requested quantities except the modulus of toughness. Estimation of the area under the stress–strain curve will be performed by approximating the area under each of the three segments of the graph. When the material is elastic, the area under the curve is simply the area of the right triangle:

```
> tough[elast] := 1/2 * ( pt2[1] - pt1[1] ) * ( pt2[2] - pt1[2] );
```

$$tough_{elast} := 4.620000000$$

which, restricted to two significant digits, is

```
> tough[elast] := evalf( tough[elast], 2 );
```

$$tough_{elast} := 4.6$$

Note that the units for the modulus of toughness are, since the strain is dimensionless, the same units as for the stress. Thus, the elastic portion of the response contributes approximately 4.6 MPa to the total modulus of toughness.

The area under the constant segment of the curve is approximated by the area of the rectangle with base 0.4% and height 330 MPa. That is,

```
> tough[const] := evalf( ( pt3[1] - pt2[1] ) * pt2[2], 2 );
```

$$tough_{const} := 1.3$$

and this segment of the curve contributes 1.3 MPa to the modulus of toughness of the material.

The area of the third section is not so easy to estimate. Lower and upper bounds on the area can be obtained by computing the area of the rectangles with the height equal to the minimum and maximum stress for strains between 3.2% and 7.4%:

```
> tough[min] := evalf( ( pt5[1] - pt3[1] ) * pt5[2], 2 ) ;
```

$$tough_{min} := 15.$$

```
> tough[max] := evalf( ( pt5[1] - pt3[1] ) * pt4[2], 2 ) ;
```

$$tough_{max} := 18.$$

Lower and upper estimates of the <u>total</u> modulus of toughness are (in MPa):

```
> Ttough[min] := evalf( tough[elast] + tough[const] + tough[min], 2 );
```

$$Ttough_{min} := 21.$$

```
> Ttough[max] := evalf( tough[elast] + tough[const] + tough[max], 2 );
```

$$Ttough_{max} := 24.$$

---

4. Refine and implement a solution

With the exception of the contribution of the third segment of the curve to the toughness of the material, all the calculations made in Step 3 are quite reasonable. To be more precise, the error in the two moduli of toughness relative to the lower estimate is

```
> Ttough[err] := ( Ttough[max] - Ttough[min] ) / Ttough[min];
```

$$Ttough_{err} := .1428571429$$

or 14%, which may not be acceptable in some applications. (Include in your version of this problem the error relative to the upper estimate.)

This error will be reduced by improving the estimation of the area under the third segment of the stress–strain curve. One way to improve the approximation of this area is to collect additional data and use more than one rectangle to approximate the area under the curve. The accuracy of the results obtained in this way is typically limited by the measurement errors. The collection of accurate data will be addressed in more detail in Chapter 4.

Another approach is to approximate this region with two trapezoids having adjacent sides at $\varepsilon = 5.3\%$, the strain at which the ultimate strength occurs. (See Problem 4 at the end of this chapter.) Yet another approach is to note that this portion of the curve closely resembles the graph of a concave-down quadratic function. The general form of stress as a quadratic function of strain is

```
> QUAD := sigma = a*epsilon^2 + b*epsilon + c;
```

$$QUAD := \sigma = a\varepsilon^2 + b\varepsilon + c$$

This function has three unknown coefficients (a, b, and c), and three distinct data points (**pt3**, **pt4**, and **pt5**) are known on this portion of the graph. The coefficients can be found as the solution of a 3×3 system of linear equations. To assemble the equations of this system, simply substitute the three data points into the quadratic function:

```
> EQ1 := subs( epsilon=pt3[1], sigma=pt3[2], QUAD );
```

$$EQ1 := 330 = .001024\,a + .032\,b + c$$

```
> EQ2 := subs( epsilon=pt4[1], sigma=pt4[2], QUAD );
```

$$EQ2 := 440 = .002809\,a + .053\,b + c$$

```
> EQ3 := subs( epsilon=pt5[1], sigma=pt5[2], QUAD );
```

$$EQ3 := 360 = .005476\,a + .074\,b + c$$

Note: To understand exactly how these commands operate, consult the online help for the **subs** command (also see Chapter 3).

The solution to this system of three linear equations is found using the **solve** command

```
> SOLN := solve( { EQ1, EQ2, EQ3 }, { a, b, c } );
```

$$SOLN := \{b = 23548.75283,\ a = -215419.5011,\ c = -202.9705215\}$$

and then normalized to two significant digits. This, however, is not appropriate. Since each data point contains only two significant digits, the coefficients in the three equations can contain only two significant digits:

```
> EQ1 := evalf( EQ1, 2 );
```

$$EQ1 := 330. = .0010\,a + .032\,b + c$$

```
> EQ2 := evalf( EQ2, 2 );
```

$$EQ2 := 440. = .0028\,a + .053\,b + c$$

```
> EQ3 := evalf( EQ3, 2 );
```

$$EQ3 := 360. = .0055\,a + .074\,b + c$$

The solution to this system is

```
> SOLN2 := solve( { EQ1, EQ2, EQ3 }, { a, b, c } );
```

$$SOLN2 := \{b = 23333.33333,\ a = -211111.1111,\ c = -205.5555556\}$$

or, with only two significant digits:

```
> SOLN2 := evalf( SOLN2, 2 );
```

$$SOLN2 := \{b = 23000.,\ a = -210000.,\ c = -210.\}$$

This leads to the following formula for the stress:

```
> stress := subs( SOLN2, rhs(QUAD) );
```

$$stress := -210000.\,\varepsilon^2 + 23000.\,\varepsilon\ -\ 210.$$

Further verification of this result can be obtained by substituting the data points **pt3**, **pt4**, and **pt5** into this equation:

```
> evalf( subs( epsilon=pt3[1], stress ), 2 );
```

$$320.$$

```
> evalf( subs( epsilon=pt4[1], stress ), 2 );
```

$$400.$$

```
> evalf( subs( epsilon=pt5[1], stress ), 2 );
```

$$290.$$

The discrepancy between the values for the stress and the values in the data points is due largely to the truncation of the coefficients in the parabola to two digits.

The area under this parabola can be approximated as a sum of areas of rectangles whose upper-right corner is a point on the curve. Maple's **student** package contains two procedures related to this concept: **rightbox** plots the function and the boxes used to approximate the area, and **rightsum** computes the approximate area of the boxes. The **value** command is needed to convert the summation returned by **rightsum** into a numeric value. (The syntax of these commands, as well as examples, is available in the online help.) These commands will be used to obtain a revised numerical approximation of this contribution to the modulus of toughness.

The first step is to load the **student** package into the Maple kernel:

```
> with( student );
```

> [D, Diff, Doubleint, Int, Limit, Lineint, Product, Sum, Tripleint,
> changevar, combine, completesquare, distance, equate, extrema,
> integrand, intercept, intparts, isolate, leftbox, leftsum, makeproc,
> maximize, middlebox, middlesum, midpoint, minimize, powsubs,
> rightbox, rightsum, showtangent, simpson, slope, trapezoid,
> value]

The **with** command returns the name of each command defined in the **student** package. The online help for *student* contains additional information about these commands.

The command of interest here is **rightbox**, which can be used as follows to produce a picture showing the approximation of the area by four rectangles.

```
> rightbox( stress, epsilon=0.032..0.074,
>               title='Right-box approximation to toughness' );
```

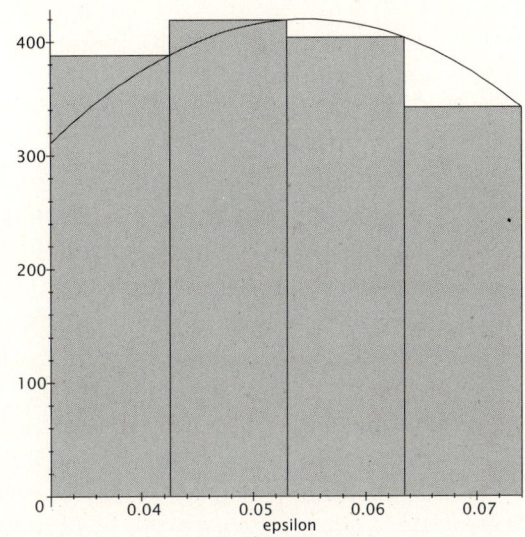

Right-box approximation to toughness

Observe that the area is overestimated in the first two rectangles and very nearly balanced by the underestimation of the last two rectangles. Thus, this four-rectangle approximation of the curve should yield a pretty good estimate of the area under the parabola. The area of these four rectangles can be written as the following sum:

```
> tough[quad4R] := rightsum( stress, epsilon=0.032..0.074 );
```

$$tough_{quad4R} := \quad .01050000000 \times$$

$$\left(\sum_{i=1}^{4} (-210000.\,(.032 + .01050000000\,i)^2 + 526. + 241.5000000\,i) \right)$$

Thus, the revised approximation of this contribution to the toughness is

```
> tough[quad4R] := evalf( tough[quad4R], 2 );
```

$$tough_{quad4R} := 18.$$

and the corresponding revised approximation of the total modulus of toughness of this material is (in MPa):

```
> Ttough[quad4R] :=
>  evalf( tough[elast] + tough[const] +tough[quad4R], 2 );
```

$$Ttough_{quad4R} := 24.$$

Note that this estimate of the toughness does fall between the maximum and minimum values found previously.

---

5. Verify and test the solution

Before you put too much confidence in this answer, you need to note all sources of errors. The impact of the measurement errors associated with the creation of the stress–strain curve and the extraction of data points from the stress–strain curve have previously been noted. Two additional approximations have been made. The portion of the stress–strain curve for $0.032 \leq \varepsilon \leq 0.074$ has been approximated by a quadratic curve, and then the area under this parabola has been approximated as the area of four rectangles.

More accurate estimates of the area under the parabolic approximation of the stress–strain curve can be obtained by replacing the rectangles with trapezoids. Note that the *See Also* list at the bottom of the online help for **rightsum** includes the command **trapezoid** (another procedure in the **student** package). The description of **trapezoid** confirms that this procedure provides the trapezoidal approximation to the area under a curve. In this case, the approximate value (again with four subintervals) is found to be

```
> tough[quad4T] :=
>    evalf( trapezoid( stress, epsilon=0.032..0.074 ), 2 );
```

$$tough_{quad4T} := 16.$$

The documentation for **rightsum** and **trapezoid** include cross-references to **simpson**. The online help confirms that this procedure returns the Simpson's rule approximation for the integral. The approximations obtained by this method are

```
> tough[quad4S] :=
>    evalf( simpson( stress, epsilon=0.032..0.074 ), 2 );
```

$$tough_{quad4S} := 17.$$

```
> Ttough[quad4S] :=
>    evalf( tough[elast] + tough[const] + tough[quad4S], 2 );
```

$$Ttough_{quad4S} := 23.$$

We now have five different approximations of the modulus of toughness for this material. Note that all five approximations are between the maximum and minimum values, and two of the estimates are identical. Although it is slightly beyond the scope of this module, Simpson's rule is exact for parabolas. Thus, this final estimate is expected to be the most accurate.

The relative error between the modulus of toughness computed using Simpson's rule and the original approximations of the toughness are

```
> error[min] := abs( Ttough[quad4S] - Ttough[min] ) / Ttough[quad4S];
```

$$error_{min} := .08695652174$$

```
> error[max] := abs( Ttough[quad4S] - Ttough[max] ) / Ttough[quad4S];
```

$$error_{max} := .04347826087$$

The corresponding error when the toughness is computed using four rectangles is

```
> error[quad4R] :=
>   abs( Ttough[quad4S] - Ttough[quad4R] ) / Ttough[quad4S];
```

$$error_{quad4R} := .04347826087$$

or 4.3%. These results suggest that estimating the toughness to be 23 MPa is accurate to approximately 5%, which is as good as can be expected for a two-significant-digit computation in this range.

What If

Suppose the material is heated. How would this change the stress–strain curve? Would the toughness increase or decrease? Why?

Four stress–strain curves are displayed in Figure 2-11. Match each curve with one of the following descriptions:

Soft and weak   Strong and tough

Weak and brittle   Hard and strong

Figure 2-11
Stress–strain curves for four different materials

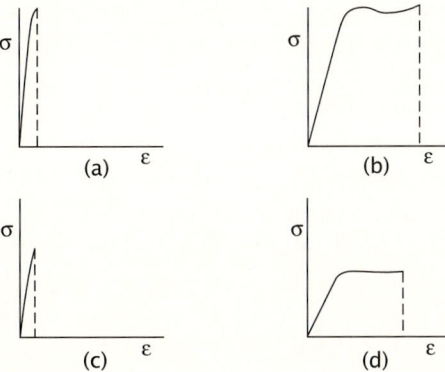

2-3 ADVANCED WORKSHEET FEATURES

The worksheet features introduced in the first section of this chapter are the fundamentals employed by all Maple users. This section introduces a few of the features that you can use to create more structured and professional-looking worksheets. For additional information and tutorials, consult the online help for the topics **worksheet[howto]** and **worksheet[glossary]** or the *Maple V Learning Guide.*

Execution Groups and Sections

Maple worksheets are built from two fundamental elements: execution groups and sections. The *execution group* is the fundamental computation and document element of a worksheet. Execution groups can be identified by the large square bracket encompassing a collection of one or more of regions. A *section* is similar, except that all sections also have a title box at the top of the element. The icon in the left margin of a section's title box controls whether the contents of the section are visible or not. The icon is a toggle switch that alternately expands and collapses the section's contents. Each execution group and section is built from a collection of input, text, and graphics regions (2D, 3D, and animation), as well as other execution groups and sections. The worksheets previously created have consisted solely of a sequence of execution groups. At this point your only exposure to sections has been in the help documents.

Try It

Open the **worksheet[howto]** help document. This worksheet contains a long collection of sections. Locate and expand the sections related to execution groups and to sections. Consult these references for detailed information about execution groups and sections.

The Execution Group selection from the Insert menu can be used to create a new execution group before or after the current cursor location. The default type for a new execution group is an input region; to convert the group to a text region, select Text Input from the Insert menu. Note that while it is possible to combine text and Maple input in the same execution group, this is strongly discouraged. (Graphics regions cannot be created by the user.)

The creation of sections is generally accomplished via the Section entry in the Insert menu. If the cursor is currently in a section, the new section will be inserted immediately after the current section; otherwise, the new section is inserted immediately following the current execution group or paragraph. A subsection is a section nested within the current element. Subsections are created via the Subsection entry in the Insert menu. The Inserting section of the **worksheet[howto]** help document contains a full description of the exact placement of a new section, subsection, or execution group.

Maple provides a collection of tools for the creation, modification, and disassembly of a section. Execution groups and sections can be merged or separated by selecting the appropriate subentry from the Split and Join option on the Edit menu. The Indent selection in the Format menu creates a subsection containing the current selection of one or more elements.

EXAMPLE 2-4

Creating Sections in a Worksheet

The worksheet *mysecond.mws* that you created earlier in this chapter (see Figure 2-5) consists of a collection of execution groups. The overall readability of the worksheet can be enhanced by the appropriate use of sections and subsections. Use the tools described previously to create a separate section for each step in the solution of the tangency problem.

SOLUTION

There is no reason to create a separate section for the header information—title, author, and date—but you should put the problem statement in its own section. To create this section, place the cursor anywhere in the region containing the problem statement, and select Indent from the Format menu. You can either cut and paste the title from the text region or manually type it in the title box (and delete it from the text region). Use this same procedure to create separate sections for the six steps in the analysis of this problem. Use subsections for the brief asides in steps 2, 3, and 6. Collapse all sections and subsections related to the solution of the problem. Reposition the cursor at the restart command, and then save the worksheet as *mythird.mws*. The worksheet should resemble Figure 2-12.

Figure 2-12
Tangency problem worksheet with sections

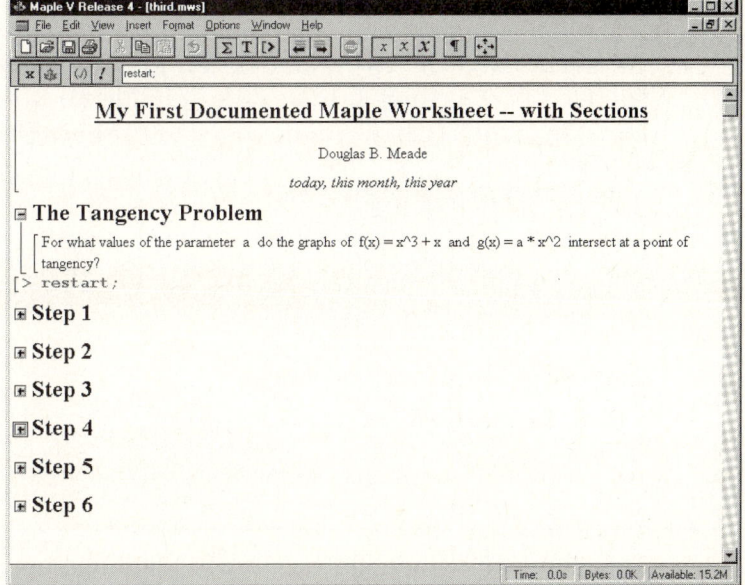

In-Line Mathematics

Any mathematical expression that can be used in a Maple command can be formatted as an *in-line mathematical expression* in a text region. There are several methods to create an in-line mathematical expression. One of the easier methods is to type the expression, using Maple syntax, in a text region. Then, after highlighting the Maple input to be reformatted, select Math Expression from the Convert To option on the Format menu. (The Format menu can also be accessed by clicking the right mouse button.) Note, from the changes in the context bar, that the current context becomes an input region (not a text region). If the string is valid Maple input, the reformatted version appears in the text region in the worksheet (see Figure 2-13). If the selection is not valid Maple input, Maple will format as much of the string as possible and insert a question mark (?) where syntax errors are detected. Note that while in-line math expressions are inert—they are not executed—Maple's formatting algorithm does apply some simplifications to the input. As a result, some simple mathematical expressions are very difficult, if not impossible, to include in a text region of a Maple worksheet.

Figure 2-13
In-line mathematical expression

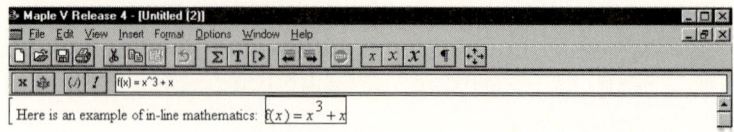

Hyperlinks

Navigation between different Maple worksheets or different sections of a single worksheet is facilitated by hyperlinks. A *hyperlink* is a segment of text, underlined and dark cyan, that, when clicked, moves the cursor to a pre-defined location in this or another worksheet. For example, the See Also sections of the help worksheets frequently contain hyperlinks to related help worksheets. To reformat a section of text as a hyperlink, highlight the text segment, and then select Hyperlink from the Convert To option on the Format menu. The hyperlink is complete once the target information is supplied in the dialog box, as shown in Figure 2-14. Additional information about hyperlinks can be found in the **worksheet[howto]** help document.

Figure 2-14
Creating a hyperlink to a help document

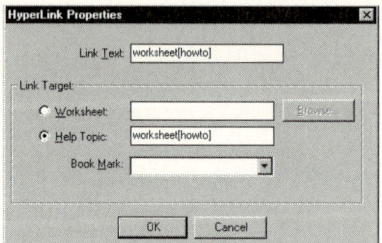

SUMMARY

This chapter introduced the Maple graphical user interface and provided an overview of the use of worksheets and online help. In learning some of Maple's capabilities, you saw a few of the commands that will be introduced in later chapters. These include **plot**, **solve**, **subs**, and **evalf**. You learned about the different region types and how to create execution groups and sections. This introduction concluded with comments about hyperlinks and the creation of in-line mathematical expressions. The general techniques, suggestions, and cautions presented here, and applied in the examples and Try It! exercises, will be used throughout the remainder of the module. A mastery of the concepts in this chapter, particularly the use of the online help, should provide a solid foundation for the continued development of your expertise with Maple.

The application in this chapter showed how a mechanical engineer can use a stress–strain curve to determine many key characteristics about a material. This application used measured data to determine the fracture point, ultimate strength, Young's modulus, and modulus of toughness and included comments on the proper use of results obtained with known accuracies. This problem will be revisited when you become more familiar with Maple commands.

Key Words

assignment	Maple worksheet
balloon help	menu bar
context bar	on-line help
current context	section
execution group	set
graphical user interface (GUI)	status line
graphics region	tangency problem
hyperlink	text region
in-line mathematical expression	title bar
input region	toolbar
list	

Maple Commands

`diff`	`simpson`
`evalf`	`solve`
`help (?)`	**student** package
`rightbox`	`subs`
`rightsum`	`trapezoid`
`simplify`	

References

1. Boresi, A.P., Schmidt, R. J., and Sidebottom, O.M. *Advanced Mechanics of Materials.* New York: John Wiley & Sons, 1993.
2. Collins, J.A. *Failure of Materials in Mechanical Design.* New York: John Wiley & Sons, 1981.
3. Heal, K.M., Hansen, M.L., and Rikard, K.M. *Maple V Learning Guide.* New York: Springer-Verlag, 1996.
4. Shigley, J.E. and Mitchell, L.D. *Mechanical Engineering Design.* New York: McGraw-Hill, 1983.

Problems

1. Repeat the graphical and symbolic verification that $a = -2$ is another solution to the tangency problem. Add these results, with appropriate documentation and explanation to the worksheet *mythird.mws*.

2. Convert all textual mathematical expressions in the worksheet created in Problem 1 to inline math expressions. Also, in a new section at

the end of the worksheet, create hyperlinks to the help documents for each of the Maple commands used in this worksheet and to the help documents most relevant to the interface features used in this worksheet. Call the resulting worksheet *myfourth.mws*.

3. Create a new worksheet, called *reference.mws*, that contains links to examples and help and other introductory material. Use hyperlinks to create links to commonly accessed online help documents, including **worksheet**, **worksheet[howto]**, **worksheet[glossary]**, **student**, and **help**. As you progress through this module, you should update this worksheet with new links, summaries of main techniques, examples, and any other information you find useful.

4. Estimate the modulus of toughness of the composite material discussed in Application 2 when the area under the third segment of the stress–strain curve is approximated by two trapezoids with a common side of $\varepsilon = 5.3\%$. What is the corresponding error in the modulus of toughness, relative to the smallest estimate of the toughness? (What benefit is obtained by using the smallest estimate in this comparison?)

5. This chapter used **rightbox** and **rightsum** to approximate the area under the quadratic curve with the default number of rectangles. How many rectangles are needed to approximate the area to four digits of accuracy? (*Hint*: This question can be answered by trial and error; consult the online help for the necessary modification to the syntax of **rightbox** and **rightsum**.)

6. Repeat the previous problem with **trapezoid** and **simpson**. What do these results tell you about the approximation errors associated with the use of rectangles, trapezoids, and quadratics to approximate the area under the parabola?

7. The integral of a nonnegative function is the limit of the area of rectangles. In the previous problems, you estimated the area under the quadratic curve to two digits of accuracy. Compute the exact area under the curve as an integral. First, use the online help to find the Maple command for computing definite integrals. How does this value compare to the approximation in this chapter and in Problem 6?

8. Investigate the use of **stats[fit]** as a means of fitting the data to a quadratic function. Verify the results in a worksheet. Then collect more data, and compute the revised fit and corresponding contribution to the modulus of toughness.

   Investigate the benefits of collecting more data and using **fit** to obtain a better approximation to the Young's modulus.

9. This chapter presented techniques for the insertion of new sections, execution groups, text regions, and other common elements of a Maple worksheet. Elements can also be removed from a worksheet. The only menu selection that relates to deletion is Delete Paragraph (under Edit). Use the online help to learn how to delete a section and execution group.

10. Create documented worksheets, including hyperlinks to relevant help documents for the solution to Example 1-1 and for the five-step problem-solving process used to analyze Application 1 in Chapter 1.

3 Engineering and Scientific Manipulations

Lift and Drag Many different types of engineers are involved in the design and manufacture of an airplane. The complexity of modern jet aircraft design requires material engineers to test exotic materials that are lightweight yet strong, using analytical methods similar to those discussed in the previous chapter. Aeronautical engineers determine the engine specifications based on the lift and drag properties of the aircraft and performance criteria, including the cruising speed and altitude, fuel efficiency, and noise restriction. Lift and drag are components of the total force on the airplane. Drag is the component of force that resists motion through the air, and lift is the component of force that keeps the plane airborne. In particular, once an airplane reaches its cruising altitude and speed, the lift balances the total weight of the aircraft. One of the engineer's jobs is to develop designs that maximize the lift-to-drag ratio.

In this chapter, you will see how an aeronautical engineer uses lift and drag to determine the thrust that an airplane's engines must generate to maintain cruising altitude and speed. Information about the lift and drag is obtained from several sources, including wind tunnel tests using scale models and mathematical models for fluid dynamics. The application in this chapter demonstrates the use of Maple to symbolically manipulate the equations that relate the variables involved in the calculation of the thrust needed to maintain cruising altitude and speed.

INTRODUCTION

Most software packages and programming languages are designed to work with numerical data and operations. These computations are almost always subject to floating-point errors. A computer algebra system, such as Maple, is distinguished by the ability to manipulate mathematical objects such as expressions, functions, equations, lists, and sets. This chapter introduces the fundamental concepts, commands, and techniques required to use Maple to enter, solve, and otherwise manipulate mathematical objects.

3-1 ASSIGNMENT AND EXPRESSIONS

The *assignment* statement provides a mechanism for assigning meaningful names to selected objects and results. The general form for a Maple assignment is **LHS := RHS**; where **LHS** is a valid Maple name and **RHS** is a Maple expression. Recall, from Chapter 1, that the semicolon is used to indicate the end of a command. The "colon equal" symbol (:=) must not be confused with the equal sign (=), which is used to form an *equation*.

A Maple *name* is usually a string, which in its simplest form is a letter followed by any combination of zero or more letters, digits, and underscores. Names are case sensitive (for example, **eqn** and **EQN** are different names). More information about names is available under the online help topic name. Examples of valid names include **expr**, **mass**, **volume**, **VARS**, **alpha**, **X**, **Theta**, **PTS**; invalid names include **5B** (names cannot begin with a number), **E%g** (% cannot be used in a name), and **in** (which is a reserved name in Maple).

Expressions are the fundamental building blocks used to construct objects in Maple. The simplest expressions are names. More complicated expressions involve combinations of names, numbers, and mathematical *operators* (such as +, *, ^).

EXAMPLE 3-1

The Mass Density of a Sphere

A solid spherical ball has mass 12 kg and volume 36 L (liters). What is the mass density of the ball?

SOLUTION

This problem, like several others discussed in this chapter, is not difficult to solve without Maple. The purpose of this example is to demonstrate how to use Maple to perform basic mathematical operations.

The first step, after Maple is reinitialized, is to assign the mass and volume to two meaningful names:

```
> restart;
> mass := 12;
```

$$mass := 12$$

```
> volume := 36;
```

$$volume := 36$$

The mass density is the quotient of the mass and volume:

```
> density := mass/volume;
```

$$density := \frac{1}{3}$$

Thus, the density is $\frac{1}{3}$ kg/L.

. . . .

Observe that Maple *substitutes* the current value, if any, of a name whenever the name appears on the right-hand side of an assignment. Following substitution, the expression is simplified. The expression that results from this simplification is then assigned to the name on the left-hand side. This automatic substitution and simplification is called *full evaluation* of an expression. For additional information about Maple's evaluation rules, see the help worksheets for **eval**, **uneval**, and **quotes**.

When Maple encounters a name that has not been assigned a value, the name stands for itself; it is an *unevaluated name*. There are a few reserved names in Maple, including **Pi** (the constant π), **I** (the square root of –1), and **infinity** (∞); a complete list of initially defined names can be found under the help topic **ininame**.

Mathematical expressions are entered in Maple using a notation similar to that used in most modern programming languages, and parentheses are used to group subexpressions. Expressions are evaluated according to the following order of precedence: ^ (or ******), *****, **/**, **+**, **-** (see **?operator** and **?operators,precedence** for the complete precedence table).

EXAMPLE 3-2

The Volume of a Sphere

The volume V of a sphere with radius R is given by $V = \frac{4 \pi R^3}{3}$. Express this relationship in Maple. Compute the volume of spheres with radii $R = 2$, $R = \sqrt{27}$, and $R = \frac{r+3}{\pi}$.

SOLUTION

To ensure that assignments from earlier work, even work done in a different worksheet, do not interfere with the current problem, you should begin the solution to this problem with the **restart** command:

```
> restart;
```

The relationship between volume and radius is

```
> V := 4/3 * Pi * R^3 ;
```

$$V := \frac{4}{3} \pi R^3$$

Note how the precedence rules are applied to obtain the desired expression to be assigned to the name *V*. To compute the volume of a sphere with radius 2, assign the value 2 to **R**:

```
> R := 2 ;
```

$$R := 2$$

and then ask Maple to display the current value of **V**:

```
> V ;
```

$$\frac{32}{3}\,\pi$$

Observe how full evaluation is used to prepare this result.

This process can be streamlined by putting both statements on the same input line. For example, for the second radius, you can enter

```
> R := sqrt(27) ; 'V' = V ;
```

$$R := 3\sqrt{3}$$
$$V = 108\,\pi\sqrt{3}$$

Observe how the output is improved by displaying an equation with an appropriately chosen unevaluated name on the left-hand side. Here, since the name **V** has already been assigned a value, the single quotation marks (') are used to prevent the full evaluation of the enclosed expression. Note also the different uses of := and =.

The volume of the third sphere is determined in the same manner:

```
> R := (r+3) / Pi ; 'V' = V ;
```

$$R := \frac{r+3}{\pi}$$

$$V = \frac{4}{3}\frac{(r+3)^3}{\pi^2}$$

Note that symbolic simplifications are made whenever possible.

. . . .

An assignment can be removed by assigning the name to itself. This is accomplished using either single quotation marks (') or the **evaln** command. Each method is illustrated in Example 3-3.

EXAMPLE 3-3

Altering Assignments

Assign the volume of the sphere with diameter *d* to the name **Vdiam**. Next, remove all assignments for the names **R** and **V**. Then find the current values assigned to **R**, **V**, and **Vdiam**.

SOLUTION

Since this problem builds upon Example 3-2, there is no reason to begin this solution with the **restart** command.

The radius of a sphere with diameter *d* is

> `R := d/2; Vdiam := V;`

$$R := \frac{1}{2} d$$

$$Vdiam := \frac{1}{6} \pi d^3$$

Single quotation marks are used to return the name **R** to its unevaluated form

> `R; R := 'R';`

$$\frac{1}{2} d$$

$$R := R$$

The **evaln** command is used to perform the same operation on the name **V**.

> `V; V := evaln(V);`

$$\frac{4}{3} \pi R^3$$

$$V := V$$

The current values of the three names used in this problem are

> `R, V, Vdiam;`

$$R, V, \frac{1}{6} \pi d^3$$

. . . .

EXAMPLE 3-4   ## Right Circular Cone

The volume and surface area of a right circular cone with height *h* and radius *r* are given, respectively, by $V = \frac{1}{3} \pi r^2 h$ and $S = \pi r^2 + \pi r \sqrt{r^2 + h^2}$. Find the volume and surface area of a cone with height 17.5 cm and radius 9.2 cm.

SOLUTION

The volume and surface area of a cone can be entered as

> `V := Pi * r^2 * h/3;`

$$V := \frac{1}{3} \pi r^2 h$$

> S := Pi * r^2 + Pi * r * sqrt(r^2+h^2);

$$S := \pi\, r^2 + \pi\, r \sqrt{r^2 + h^2}$$

Using the radius and height given in the problem,

> r := 17.5; h := 9.2;

$$r := 17.5$$
$$h := 9.2$$

the volume (in cm^3) and surface area (in cm^2) are

> volume = V, surface_area = S;

$$volume = 939.1666666\,\pi,\ surface\_area = 652.2414197\,\pi$$

. . . .

Try It

Express, for a general radius, the volume and surface area of a cone whose height is twice the radius.

EXAMPLE 3-5

assume **and** about

Notice that Maple does not automatically simplify $\sqrt{r^2}$ to r. (Why is this not, in general, a valid simplification?) The user can provide additional information to Maple in a number of ways. One of the most sophisticated and flexible methods is Maple's *assume facility*. For example, for a geometric problem such as the ones considered in this section, you know the radius is never negative, that is, $r \geq 0$.

SOLUTION

Maple does not automatically replace $\sqrt{r^2}$ with r because this simplification is correct only when r is nonnegative. For example,

> restart;
> sqrt(3^2), sqrt((-3)^2);

$$3, 3$$

The **assume** command can be used, as follows, to provide additional information to Maple:

> assume(r>=0);

A complete list of information about any Maple name can be obtained with the **about** command:

```
> about( r );
```

Originally r, renamed r~:
is assumed to be: RealRange(0,infinity)

Notice that Maple appends a tilde (~) to a name whenever additional assumptions have been provided. With this information, Maple is able to make more progress in the simplification of $\sqrt{r^2}$.

```
> sqrt( r^2 );
```

$$r\sim$$

Assumptions remain in effect until the name is returned to its unevaluated state:

```
> r := 'r';
```

$$r := r$$

```
> about( r );
```

r:
nothing known about this object

3-2 EXPRESSION SEQUENCES, LISTS, AND SETS

The last output in each of Examples 3-3 and 3-4 is an *expression sequence*, that is, a collection of Maple expressions separated by commas. Expression sequences are used in many different contexts, including sets, lists, arguments to functions of more than one variable, and indices for tables and arrays. Some of the uses of expression sequences will be considered in this section; additional information can be found in the online documentation for the keyword **exprseq**.

A *list* is an ordered expression sequence enclosed in square brackets (**[ ]**). A *set* is an unordered expression sequence enclosed in curly braces (**{ }**).

EXAMPLE 3-6

Lists and Sets

Construct a list and a set from the expression sequence

```
> EXPRSEQ :=1, 2, 2+1, 2^2-log(exp(2)), 3*density, volume,
>          Pi*radius^2, volume:
```

Compare the elements of the list and set and the order in which the individual terms appear.

SOLUTION

On the surface, the only difference between a list and set is the delimiter of the expression sequence.

```
> LIST := [ EXPRSEQ ];
```

$$LIST := [1, 2, 3, 2, 3 \text{ } density, \text{ } volume, \pi \text{ } radius^2, \text{ } volume]$$

```
> SET := { EXPRSEQ };
```

$$SET := \{\pi \text{ } radius^2, \text{ } volume, 3 \text{ } density, 1, 2, 3\}$$

In practice, however, the results are quite different. In particular, no element appears more than once in the set, and the elements of the set are not in the same order as the expression sequence **EXPRSEQ**. In this way, a Maple set is exactly like the sets used in mathematics. (The only difference is that Maple cannot work with sets that have an infinite number of elements.)

Set operations available in Maple include **union**, **intersect**, **minus**, and **member**.

EXAMPLE 3-7

union, intersect, and minus

Construct the union, intersection, and difference of the set of prime numbers less than 10 (that is, 2, 3, 5, 7) and the set of odd integers less than 10 (that is, 1, 3, 5, 7, 9).

SOLUTION

The sets of prime and odd integers less than 10 are

```
> PRIME := { 2, 3, 5, 7 }:
> ODD := { 1, 3, 5, 7, 9 }:
```

The union and intersection of these sets are

```
> ODDorPRIME := ODD union PRIME;
```

$$ODDorPRIME := \{1, 2, 3, 5, 7, 9\}$$

```
> ODDandPRIME := ODD intersect PRIME;
```

$$ODDandPRIME := \{3, 5, 7\}$$

Set differences are not commutative. The set of odd integers that are not primes is

```
> ODDnotPRIME := ODD minus PRIME;
```

$$ODDnotPRIME := \{1, 9\}$$

and the set of primes that are not odd is

```
> PRIMEnotODD := PRIME minus ODD;
```

$$PRIMEnotODD := \{2\}$$

. . . .

Individual elements of a list can be referenced by their position in the list. The second element of the list **LIST** can be referenced as either **LIST[2]** or **op(2,LIST)**. Consecutive elements of a list can be accessed by specifying a range of indices. The number of elements in a list (or set) is obtained by **nops**. For example, **LIST[2..nops(LIST)-1]** returns the list formed by deleting the first and last elements of **LIST**. Note that though it is permitted to refer to the second element of a set, the results can be unpredictable since the order of the elements in a set cannot be controlled by the user.

Some sequences, such as the terms in a geometric sequence, can be constructed using an explicit formula inside a loop. Commands for the construction of expression sequences, lists, and sets include the **seq** command and **$** operator. The empty expression sequence is **NULL**. The empty set is **{ NULL }** or **{ }**; the empty list is **[NULL]** or **[ ]**.

| **EXAMPLE 3-8** | ## Working with Large Sets |

Find all integers up to and including one million that are both a perfect square and a perfect cube of an integer.

SOLUTION

The set, **SQR**, of all perfect squares that do not exceed one million contains 1000 elements. These elements would be almost impossible to list by hand. The **seq** command provides a very simple means to create this set:

```
> SQR := { seq( i^2, i=1..1000 ) }:
```

Similarly, the set, **CUB**, of all perfect cubes up to 1,000,000 contains 100 elements. This set could be created using **seq**; the same result can also be obtained using the dollar (**$**) operator:

```
> CUB := { i^3 $ i=1..100 }:
```

The integers that are both perfect squares and perfect cubes that are less than a million are given by

```
> ANS := SQR intersect CUB;
```

$$ANS := \{64, 729, 1000000, 531441, 262144, 4096, 15625, 1, 46656, 117649\}$$

Note that there is no order to the elements of the set **ANS**. To organize these numbers into increasing order, the set needs to be converted into a list, which can then be sorted, that is,

```
> sort( convert( ANS, list ) );
```

$$[1, 64, 729, 4096, 15625, 46656, 117649, 262144, 531441, 1000000]$$

The number of integers that do not exceed 1,000,000 that are both perfect squares and perfect cubes is

> **nops(ANS);**

$$10$$

. . . .

Try It
◆

Suppose you want to digitize an analog voice signal, which ranges from 0 mVolts to 50 mVolts in such a manner as to use binary bits (zeros or ones). You decide that quantizing the amplitude level into 128 discrete and equal-width intervals over the range of 0 to 50 mVolts will be sufficient. Use the **seq** command to generate a list of the 128 levels that will be represented by these binary codes.

3-3 CREATION AND DISSECTION OF EQUATIONS

An *equation* is an expression containing a left-hand side (**LHS**) and a right-hand side (**RHS**) connected with an equal sign: **LHS = RHS**, where **LHS** and **RHS** are Maple expressions. Note the distinction between the use of = (for equations) and := (for assignments). An equation was used to improve the formatting of the output in Example 3-4. The expressions that form the two sides of the equation can be accessed with the **lhs** and **rhs** commands.

EXAMPLE 3-9

An Equation for a Line

One equation for the straight line through the points (*x0,y0*) and (*x1,y1*) is
$\dfrac{y - y0}{x - x0} = \dfrac{y1 - y0}{x1 - x0}$. Enter this equation in Maple, then use **rhs** and **lhs** to write the equivalent equation in which the right-hand side is 1.

SOLUTION
The original equation of the line through the points (*x0,y0*) and (*x1,y1*), where $x0 \neq x1$, is

> **LINE := (y-y0)/(x-x0) = (y1-y0)/(x1-x0);**

$$LINE := \frac{y - y0}{x - x0} = \frac{y1 - y0}{x1 - x0}$$

To rewrite this equation with a right-hand side of 1, simply divide both sides of the original equation by its right-hand side:

> **LINE1 := lhs(LINE)/rhs(LINE)=1;**

$$LINE1 := \frac{(y - y0)(x1 - x0)}{(x - x0)(y1 - y0)} = 1$$

. . . .

Try It

The equations found in Example 3-9 are not defined for certain combinations of the points (*x0,y0*) and (*x1,y1*). Use Maple to manipulate **LINE** into an equivalent form that does not involve fractions.

(*Hint:* The numerator and denominator of a fraction can be accessed via the **numer** and **denom** functions. Use the online help to determine how to use **numer** and **denom**.)

Note that an *inequality* is essentially the same as an equation, except that the equal sign is replaced with one of the inequality signs: <, <=, <>, >, >=.

3-4 SOLVING EQUATIONS AND SYSTEMS OF EQUATIONS

One of the most common uses of Maple is to solve equations. For this purpose, Maple provides the **solve** command. The syntax of the **solve** command closely parallels the manner in which we would state the problem of solving an equation for a specific variable. That is, **solve** needs to know both the equation to be solved and the variable that is to be solved for. For example,

```
> solve( a*x=b, x );
```

$$\frac{b}{a}$$

If the second argument is a set, then the output from **solve** will appear as a set containing an equation

```
> solve( a*x=b, { x } );
```

$$\{x = \frac{b}{a}\}$$

The latter form is required for a system of equations and is often preferred for a single equation because of the extra information that it contains.

EXAMPLE 3-10

Slope-Intercept Equation of a Line

Express the equation of the straight line through the points (*x0, y0*) and (*x1, y1*) in the slope-intercept form $y = mx + b$. Identify both the slope *m* and *y*-intercept *b*.

SOLUTION

The equation of this line has been previously assigned to **LINE**:

```
> LINE;
```

$$\frac{y - y0}{x - x0} = \frac{y1 - y0}{x1 - x0}$$

To put this equation in slope-intercept form, it is first necessary to solve the equation **LINE** for *y*:

```
> SOLy := solve( LINE, { y } );
```

$$SOLy := \{ y = -\frac{y0\ x1 + y1\ x - y1\ x0 - y0\ x}{-x1 + x0} \}$$

The fact that this is a set is irrelevant. If this is bothersome, simply use **op** to extract the contents of the set and thereby eliminate the curly brackets:

```
> LINEa := op( SOLy );
```

$$LINEa := y = -\frac{y0\ x1 + y1\ x - y1\ x0 - y0\ x}{-x1 + x0}$$

The slope and intercept are still somewhat hidden in this equation. What is needed is to collect all the terms involving *x* and all the constant terms. The **collect** function will do precisely this:

```
> collect( LINEa, x );
```

$$y = -\frac{(y1 - y0)\ x}{-x1 + x0} - \frac{y0\ x1 - y1\ x0}{-x1 + x0}$$

It is now a matter of inspection to see that the slope is $m = \dfrac{y1 - y0}{x1 - x0}$ and

the intercept is $b = \dfrac{y0\ x1 - x0\ y1}{x1 - x0}$.

. . . .

Some equations (for example, polynomials of a degree higher than one) have more than one solution. Often, Maple is able to find all the roots of the equation. In these cases, the output from **solve** will be an expression sequence. A specific solution can then be selected using one of the techniques discussed in Section 3-2.

Try It   The quadratic equation $ax^2 + bx + c = 0$ has two solutions. Use **solve** to find these solutions, and then verify that these solutions are consistent with the quadratic formula. When $a > 0$, $b > 0$, and $c < 0$, the discriminant $b^2 - 4ac$ is positive and larger than b^2. Thus, there will be one positive root and one negative root. Assign the positive root to the name **POS** and the negative root to the name **NEG**.

EXAMPLE 3-11   ## Intersection of a Line and a Circle

Find all points of intersection between the line $x + y = 1$ and the circle $x^2 + y^2 = r^2$.

SOLUTION

The line and circle can be specified as

> `LINE := x+y=1;`

$$LINE := x + y = 1$$

> `CIRCLE := x^2 + y^2 = r^2;`

$$CIRCLE := x^2 + y^2 = r^2$$

Points of intersection occur at points (x,y) where both equations are simultaneously satisfied. The **solve** command can be used to find these points. The only difference from our earlier use is that the first argument is the set of two equations, and the second argument is the set of two variables:

> `SYS := { LINE, CIRCLE };`

$$SYS := \{x^2 + y^2 = r^2, x + y = 1\}$$

> `VARS := { x, y };`

$$VARS := \{x, y\}$$

> `SOL := solve( SYS, VARS );`

$$SOL := \{y = \text{RootOf}(2\,\_Z^2 - 2\,\_Z + 1 - r^2),$$
$$x = -\text{RootOf}(2\,\_Z^2 - 2\,\_Z + 1 - r^2) + 1\}$$

The appearance of **RootOf** in this result is typical for polynomial equations. The **allvalues** command can be used to obtain explicit roots from **RootOf**:

> `SOL := [ allvalues( SOL ) ];`

$$SOL := \left[\{x = \frac{1}{2} - \frac{1}{2}\sqrt{-1 + 2\,r^2}, y = \frac{1}{2} + \frac{1}{2}\sqrt{-1 + 2\,r^2}\}, \right.$$
$$\left. \{x = \frac{1}{2} + \frac{1}{2}\sqrt{-1 + 2\,r^2}, y = \frac{1}{2} - \frac{1}{2}\sqrt{-1 + 2\,r^2}\} \right]$$

Observe that Maple has found two points of intersection between the curve and line. Is this always true?

. . . .

3-5 SUBSTITUTION AND EVALUATION

One method of checking your solution to the Try It! exercise in Section 3-4 is to evaluate the two roots after assigning specific values to the constants a, b, and c. However, you will need to unassign (using either **evaln** or single quotation marks) the constants prior to doing any further symbolic processing with these roots.

EXAMPLE 3-12   ## Symbolize Roots of a Quadratic Equation

Find the two roots of the general quadratic equation ($ax^2 + bx + c = 0$) for the three cases *i*) $c = 0$, *ii*) $b = 2$, and *iii*) $c = 0$ and $b = 2$.

SOLUTION

The general quadratic equation is

```
> EQN := a*x^2 + b*x + c = 0 ;
```

$$EQN := a^2 x + bx + c = 0$$

The roots of the quadratic equation are

```
> ROOTS := [ solve( EQN, x ) ];
```

$$ROOTS := \left[\frac{1}{2} \frac{-b + \sqrt{b^2 - 4\,a\,c}}{a}, \frac{1}{2} \frac{-b - \sqrt{b^2 - 4\,a\,c}}{a} \right]$$

Note the way that square brackets are used to create a list from the expression sequence returned by the **solve** command. One advantage of a list over a set is that the order of the elements in a list is fixed and can thereby be referred to uniquely by their subscripted element positions (elements of a set have no well-defined order).

To see the roots when $c = 0$, assign the value zero to **c** and look at the roots.

```
> c := 0; EQN; ROOTS;
```

$$c := 0$$
$$a\,x^2 + b\,x = 0$$
$$\left[\frac{1}{2} \frac{-b + \sqrt{b^2}}{a}, \frac{1}{2} \frac{-b - \sqrt{b^2}}{a} \right]$$

Notice that $\sqrt{b^2}$ is not automatically simplified to b. Unlike in Example 3-5, there is no additional information that might allow Maple to simplify this result. However, from the observation that $\sqrt{b^2} = -b$ when $b < 0$, you can conclude that $\sqrt{b^2} = |b|$, for any real values of b. (Since Maple makes simplifications that are valid for all real and complex numbers, $\sqrt{b^2}$ is not replaced with $|b|$; see **?sqrt** for additional details.)

The expression for the roots when $b = 2$ should be obtainable in the same manner:

```
> b := 2; EQN; ROOTS;
```

$$b := 2$$
$$a\,x^2 + 2\,x = 0$$
$$\left[\frac{1}{2} \frac{-2 + \sqrt{4}}{a}, \frac{1}{2} \frac{-2 - \sqrt{4}}{a} \right]$$

Note, however, that these are the roots when both $b = 2$ and $c = 0$. The **simplify** command can be used to further simplify the two roots of $ax^2 + 2x$:

```
> simplify( ROOTS );
```

$$\left[0, -\frac{2}{a} \right]$$

To examine the roots when just $b = 2$, the assignment to **c** must be removed:

```
> c := 'c';
```

$$c := c$$

Prior to examining the roots, it is a good idea to verify that all three constants evaluate as you expect:

```
> a,b,c;
```

$$a, 2, c$$

Good. The roots to the quadratic equation when $b = 2$ are

```
> ROOTS;
```

$$\left[\frac{1}{2} \frac{-2 + \sqrt{4 - 4\,a\,c}}{a}, \frac{1}{2} \frac{-2 - \sqrt{4 - 4\,a\,c}}{a} \right]$$

which simplify to

```
> simplify( ROOTS );
```

$$\left[\frac{-1 + \sqrt{1 - a\,c}}{a}, -\frac{1 + \sqrt{1 - a\,c}}{a} \right]$$

. . . .

Try It Determine when the two roots of the quadratic equation will differ by a constant Δ. When are the roots equal ($\Delta = 0$)?

The recurring need to unassign variables can be annoying. The **subs** command provides an alternative method for substituting values into an equation without making assignments to the names. The **subs** command accepts, in its simplest form, two arguments. The first argument, which must be an equation or set of equations, specifies the substitution(s) to be applied to the second argument. Substitution does not include full evaluation of the resulting expression; use **simplify** to force full evaluation.

EXAMPLE 3-13   ## Using subs to Avoid Assignments

Repeat Example 3-12 using **subs**.

SOLUTION

The first step is to remove all previous assignments to **a**, **b**, and **c**. (Note that **restart** could have been used, but then it would be necessary to recompute **ROOTS**.)

```
> a := 'a': b := 'b': c := 'c':
> a,b,c;
```

$$a, b, c$$

The quadratic equation with $c = 0$ is found, using **subs**, to be

```
> subs( c=0, EQN );
```

$$a\,x^2 + b\,x = 0$$

The corresponding roots to this equation are

```
> subs( c=0, ROOTS );
```

$$\left[\frac{1}{2} \frac{-b + \sqrt{b^2}}{a}, \frac{1}{2} \frac{-b - \sqrt{b^2}}{a} \right]$$

In the same way, the quadratic equation with $b = 2$ and the corresponding roots are

```
> subs( b=2, EQN );
```

$$ax^2 + 2x + c = 0$$

```
> subs( b=2, ROOTS );
```

$$\left[\frac{1}{2} \frac{-2 + \sqrt{4 - 4\,a\,c}}{a}, \frac{1}{2} \frac{-2 - \sqrt{4 - 4\,a\,c}}{a} \right]$$

Since **subs** does not automatically simplify its results, it is often necessary to explicitly request this simplification:

```
> simplify(" );
```

$$\left[\frac{-1 + \sqrt{1 - a\,c}}{a}, -\frac{1 + \sqrt{1 - a\,c}}{a} \right]$$

The case where $b = 2$ and $c = 0$ can be handled by specifying all substitutions in a set:

```
> SUBS := { b=2, c=0 }:
> subs( SUBS, EQN );
```

$$ax^2 + 2x = 0$$

```
> simplify( subs( SUBS, ROOTS ) );
```

$$\left[0, -\frac{2}{a} \right]$$

Observe that these results are in perfect agreement with Example 3-12. Moreover, note that the three coefficients still evaluate to their names:

```
> a,b,c;
```

$$a, b, c$$

. . . .

Note the way a set is used to specify multiple substitutions in the last part of Example 3-13. In this case, the substitutions are applied simultaneously. If the two substitutions were instead specified as an expression sequence, then the substitutions would be applied in succession, from left to right, to the last argument of **subs**.

Try It  Predict, and explain, the results produced by the following two commands:

```
> TEST1 := subs( { x = y, y = x }, [x,y] ):
> TEST2 := subs( x=y, y=x, [x,y] ):
```

(*Hint:* Consult the help worksheet for **subs**.)

Application 3   **LIFT AND DRAG**

One of the fundamental jobs of an aeronautical engineer is the design and analysis of airplanes. In this application, you will learn about lift and drag and use this information to determine the thrust required to balance the drag on an airplane cruising at a fixed altitude. The biggest complication in this problem is the number of variables and equations that have to be managed. The solution basically amounts to using algebra to manipulate the relations between the parameters to a form where they can be substituted into another expression. The use of experimental data can also be problematic; this is addressed in more detail in Chapter 4.

Fundamentals

The parameters involved in the description of the forces acting on an airplane can be divided into two categories. The first group describes characteristics of the airplane, including the *surface area, S,* of the wing of the plane, the *wing span, b,* and the plane's *velocity, V* (in feet per second). The second group of parameters describes the environment through which the airplane travels. Included in this group are the *air density* (in slugs per cubic foot) at flight altitude, ρ, and at sea level, ρ_{SL}, the *air pressure* (in pounds per square inch, psi) at flight altitude, p, and sea level, p_{SL}, and the speed of sound (in ft/sec) at flight altitude, a.

In addition to these physical parameters, several dimensionless quantities will be used in the analysis. The *Mach number*, $M = V/a$, is the dimensionless quantity that normalizes the plane's velocity with respect to the speed of sound. Similarly, you will need to define $\delta = p/p_{SL}$ and $\sigma = \rho/\rho_{SL}$. Also, the ratio of the specific heats at constant pressure and constant volume is the dimensionless parameter $\gamma = \rho a^2/p$, which is known to be approximately 1.4 for air. The final dimensionless quantity is the *aspect ratio*, $AR = b^2/S$, where S is measured in square feet and b in feet.

Lift, L, and *drag*, D, are two components of the total force on the plane. In terms of the quantities given above,

$$L = \frac{\rho\,V^2}{2}\ S\ C_L \text{ and } D = \frac{\rho\,V^2}{2}\ S\ C_D.$$

Furthermore, the dimensionless coefficients of lift and drag, C_L and C_D, are related by $C_D = C_{D_0} + \alpha C_L^2$ for appropriate constants C_{D_0} and α. The *thrust* is the force of propulsion provided by the aircraft's engines. When the plane is flying at a constant altitude and speed, the thrust must exactly balance the drag.

Suppose the plane is flying at an altitude of 35,000 feet at Mach 0.84, the wingspan is $b = 200$ ft, the aspect ratio is $AR = 10$, and the plane's total weight (including craft, passengers, cargo, fuel) is 500,000 pounds. Air pressure and density at sea level are $p_{SL} = 14.7$ psi and $\rho_{SL} = 0.002378$ slugs/ft^3 and, at 35,000 feet, $\delta = 0.2351$ and $\sigma = 0.3096$.

The wind tunnel research team and the computational fluid dynamics (CFD) team have supplied data indicating the relationship between the lift and drag coefficients (see Table 3-1). The wind tunnel data was obtained experimentally from a scale model of the airplane. The CFD data is based on computational tests that use numerical methods to solve partial differential equations that have been developed to model the forces acting on the airplane.

Note: Compressibility effects, which become more significant as the Mach number increases toward Mach 1, have not been fully accounted for in this analysis.

Table 3-1 Experimental Lift and Drag Data.

Wind Tunnel Data		CFD Data	
C_L	C_D	C_L	C_D
0.01	0.0155	0.00	0.0157
0.13	0.0165	0.11	0.0163
0.21	0.0181	0.25	0.0187
0.40	0.0249	0.38	0.0228
0.65	0.0380	0.51	0.0293
		0.66	0.0389
		0.80	0.0501

Comment on Consistent Unit Systems

Units can be a source of confusion. Newton's Second Law of Motion, $F = ma$, often serves as the basis for defining units, as well as converting from one system to another.

In the *mks*, or *SI (System International)*, system, a force of one newton accelerates a mass of one kilogram one meter per second squared, that is, $1 \text{ N} = 1 \text{ kg m/sec}^2$. The fact that these units satisfy Newton's Second Law of Motion means that this system of units is consistent. Another metric system that is consistent is the *cgs* system, where a force of 1 dyne accelerates a mass of one gram by one cm/sec^2. From this, it is clear that 100,000 dynes is equivalent to 1 newton (N).

In many fields, English systems of units are still used; some of these systems are consistent, and others are not. An additional confusion with English systems is the fact that mass and force (weight) often have the same units (pounds). In fact, on earth, a mass of 1 pound (lb_m) weighs 1 pound (lb_f). Consistency with Newton's Second Law of Motion is preserved for this system only if pounds mass (lb_m) are converted to the mass unit known as slugs. This conversion is accomplished by dividing the slugs by the acceleration due to gravity, 32.174 ft/sec^2. Thus, a mass of 1 slug would accelerate 1 ft/sec^2 if subjected to a force of 1 pound (lb_f). The consistent system formed by feet, slugs, and seconds is known as the *English Gravitational System*. The system that uses feet, pounds (mass), and seconds, known as the *English Engineering System*, is not consistent since a force of 1 pound (lb_f) accelerates a mass of 1 pound (lb_m) 32.174 ft/sec^2. The *Absolute English System*, where a force of 1 poundal accelerates a mass of 1 pound (lb_m) 1 ft/sec^2, is consistent, and 1 poundal = $1/32.174 \text{ lb}_m = 0.03108 \text{ lb}_m$. The bottom line is that when working with English units, either convert mass to slugs (1 slug=32.174 lbm) or convert force (lb_m) to poundal.

A newton is a smaller amount of force than a pound (lb_f): 4.448 newton = 1 lb_f. Since a slug of mass would be accelerated 1 ft/sec^2 by 1 lb_f of force, it follows that 1 newton of force would accelerate 1 slug 0.2248 ft/sec^2, or would accelerate 0.2248 slug 1 ft/sec^2, or $0.2248 \times 0.3048 \text{ m/sec}^2$, giving us 0.0685 slug = 1 kg, or 14.59 kg = 1 slug.

To illustrate, the fuel efficiency and range of an airplane are related to the Thrust Specific Fuel Consumption (TSFC) parameter. Conversion of a TSFC parameter of 0.75 lb_f/lb_m-hr (which has units of 1/speed) to a consistent set of units requires the conversion of mass from lb_m to slug or the conversion of the force from lb_f to poundal, in either case necessitating dividing TSFC by 32.174 ft/sec^2. In addition, it is necessary to use the fact that there are 3600 seconds per hour to convert hours to seconds. The final result is a TSFC = $0.75 \times 8.634 \times 10^{-6}$ lb_f/slug-sec in the (consistent) English Gravitational System. We will use this particular result in Problem 6.

◢ 1. Define the problem

The ultimate objective of this application is to determine the thrust necessary to balance the drag at cruise conditions.

◢ 2. Gather information

The cruising altitude for this plane is given as 35,000 feet at Mach 0.84. The corresponding air pressure and density parameters are found in atmospheric tables: $p_{SL} = 14.696$ lb/in^2 and $\delta_{35000} = 0.2360$ while $\rho_{SL} = 0.002377$ slugs/ft^3; $\sigma_{35000} = 0.3106$. Also, $\gamma = 1.4$. You are given the following structural data:

Weight:	500,000 lb
Wing span:	200 ft
Aspect ratio:	10.0

Experimental values for the coefficients of lift and drag are given in Table 3-1. More generally, the coefficients of lift and drag are related by the formula $C_D = C_{D_0} + \alpha C_L^2$. The actual forces of lift and drag are given

by $L = \dfrac{\rho V^2}{2} S\ C_L$ and $D = \dfrac{\rho V^2}{2} S\ C_D$.

Note that the problem involves a combination of physical and dimensionless parameters. Conversions between the different parameters are

made with the use of the formulas $AR = {b^2}/{S}$, $M = {V}/{a}$, $\gamma = {\rho a^2}/{p}$, $\delta = {p}/{p_{SL}}$,

$\sigma = {\rho}/{\rho_{SL}}$.

◢ 3. Generate and evaluate potential solutions

The successful solution of this problem depends on the development and implementation of a good strategy.

You should begin with the following observations. Under cruising conditions the plane is flying in a level path at a constant speed. That is, the lift component of the force exactly balances the weight of the plane

```
> restart;
> balance1 := lift = weight;
```

$$balance1 := lift = weight$$

and the thrust from the engines matches the drag force

```
> balance2 := thrust = drag;
```

$$balance2 := thrust = drag$$

This suggests the following three-stage plan of attack. Since the weight is known, the coefficient of lift, C_L, can be determined. Either the experimental data or the explicit formula relating the coefficients of lift and drag will be used to find the corresponding drag coefficient, C_D. The drag force, which must also be the thrust, can then be directly computed.

The numerical parameters pertaining to this problem are collected in a single list. The names **p0** and **rho0** denote the pressure and density at sea level, and **gamma1** is used because **gamma** is a reserved name in Maple. Once the air pressure at sea level is converted from lb/in^2 to lb/ft^2, all units are in the English Gravitational System. Note that four significant digits are provided for all parameter values.

```
> PARAM := evalf( [ w = 500000, b = 200, AR = 10, M = 0.84,
>                   gamma1 = 1.4, p0 = 14.696*12^2, delta = 0.2360,
>                   rho0 = 0.002377, sigma = 0.3106 ], 4 );
```

$$PARAM := [w = 500000., b = 200., AR = 10., M = .84, gamma1 = 1.4,$$
$$p0 = 2116., \delta = .2360, rho0 = .002377, \sigma = .3106]$$

Since the lift is given by

```
> lift := rho*V^2/2 * S * CL;
```

$$lift := \frac{1}{2} \rho \, V^2 \, SCL$$

the balance of lift and weight can be expressed as

```
> balance1;
```

$$\frac{1}{2} \rho \, V^2 \, S \, CL = weight$$

and the coefficient of lift is, in general:

```
> coefL := op( solve( balance1, { CL } ) );
```

$$coefL := CL = 2 \, \frac{weight}{\rho \, V^2 \, S}$$

This shows that C_L is a function of weight, density, velocity, and surface area. The weight is explicitly given; the other three quantities must be computed from the given data and the relations given at the end of Step 2. As a first step, note that

```
> VARS := [ weight=w, V = M*a, S = b^2/AR, rho=sigma*rho0 ];
```

$$VARS := \left[weight = w, \; V = M \, a, \; S = \frac{b^2}{AR}, \; \rho = \sigma \, \rho 0 \right]$$

Observe that the speed of sound is not yet known, so substituting the parameters into **VARS** does not give the plane's velocity:

```
> subs( PARAM, VARS );
```

$$[weight = 500000., \; V = .84 \, a, \; S = 4000.000000, \; \rho = .0007382962]$$

The speed of sound, a, can be expressed in terms of the pressure and density at sea level and the dimensionless constants δ, σ, and γ. The general formula for the speed of sound (in ft/sec) is

```
> Vsound := subs( [ p=delta*p0, rho=sigma*rho0 ],
>                        a=sqrt(p/rho*gamma1) );
```

$$Vsound := a = \sqrt{\frac{\delta \, p0 \, \gamma 1}{\sigma \, \rho 0}}$$

which yields, for this problem,

```
> subs( PARAM, Vsound );
```

$$a = 972.7904015$$

Thus, the speed of sound is $a = 972.8$ ft/sec or 663.3 mi/hr. General expressions for the four variables involved in the computation of C_L are now seen to be

```
> VARS := subs( Vsound, VARS );
```

$$VARS := \left[weight = w, \; V = M \sqrt{\frac{\delta \, p0 \, \gamma 1}{\sigma \, \rho 0}}, \; S = \frac{b^2}{AR}, \; \rho = \sigma \, \rho 0 \right]$$

The specific values, with four significant digits, for this problem are

```
> VARS2 := evalf( subs( PARAM, VARS ), 4 );
```

$$VARS2 := [weight = 500000., \; V = 817.5, \; S = 4000., \; \rho = .0007383]$$

The corresponding coefficient of lift is

```
> coefL2 := evalf( subs( VARS2, coefL ), 4 );
```

$$coefL2 := CL = .5066$$

This completes the first stage of the solution.

A quick glance at Table 3-1 reveals that the CFD tests found that the coefficient of drag corresponding to $C_L = .5066$ is approximately $C_D = 0.0293$.

> **coefD := CD = 0.0293;**

$$coefD := CD = .0293$$

Note that this data is accurate to no more than three significant digits. As a result, all subsequent computations will also have this reduced number of significant digits.

The final stage is to compute the thrust that balances the drag force on the airplane. The general formula for the drag is

> **drag := rho*V^2/2 * S * CD;**

$$drag := \frac{1}{2}\rho\,V^2 S\,CD$$

The numerical values for ρ, V, and S found in stage 1 are still valid. Once the coefficient of drag is added to the list of variables

> **VARS3 := [op(VARS2), coefD];**

$$VARS3 := [weight = 500000., V = 817.5, S = 4000., \rho = .0007382, CD = .0293]$$

the thrust requirement, accurate to three significant digits, is found to be

> **evalf(subs(VARS3, balance2), 3);**

$$thrust = 28900.$$

That is, the engines need to produce 28,900 pounds (lb_f) of thrust.

4. Refine and implement a solution

The fact that the coefficient of lift appeared in the experimental data was fortunate and completely coincidental. If the CFD data were not available, the wind tunnel data would have to be used to determine the drag coefficient corresponding to $C_L=0.5060$. One way in which this could be accomplished is to use two of the wind tunnel data points to solve for the unknown coefficients in the lift-to-drag equation.

> **liftdrag := CD = CD0 + alpha*CL^2;**

$$liftdrag := CD = CD0 + \alpha\,CL^2$$

Any pair of wind tunnel data points could be used to determine C_{D_0} and α. Using the two points with lift coefficients closest to $C_L=0.5060$ leads to the two equations:

> **eq1 := subs([CL=0.40, CD=0.0249], liftdrag);**

$$eq1 := .0249 = CD0 + .1600\alpha$$

> **eq2 := subs([CL=0.65, CD=0.0380], liftdrag);**

$$eq2 := .0380 = CD0 + .4225\alpha$$

The solution to this system of two equations for the two coefficients is

```
> LDcoef := solve( { eq1, eq2 }, { CD0, alpha } );
```

$$LDcoef := \{CD0 = .01691523810, \alpha = .04990476190\}$$

The corresponding relationship between lift and drag is

```
> LtoD := evalf( subs( LDcoef, liftdrag ), 3 );
```

$$LtoD := CD = .0169 + .0499\, CL^2$$

When the lift coefficient is $C_L = 0.5060$, this formula predicts the drag coefficient will be

```
> coefD2 := evalf( subs( CL=0.5060, LtoD ), 3 );
```

$$coefD2 := CD = .0297$$

Since this drag coefficient is slightly larger than the drag coefficient found from the CFD data, additional thrust will be needed:

```
> VARS4 := [ op( VARS2 ), coefD2 ]:
> evalf( subs( VARS4, balance2 ), 3 );
```

$$thrust = 29300.$$

5. Verify and test the solution

The computations of drag coefficient in Steps 3 and 4 have not made full use of the experimental data. Data-fitting techniques that use all the data are introduced in Section 4-4. Problem 7 in Chapter 4 finds, for both the wind tunnel and CFD data, a lift-to-drag function that uses all the data points. These drag coefficients are between the two values found in Steps 3 and 4. Thus, all the drag coefficients differ by no more than 2.1 percent. The thrust projections exhibit the same relative error since drag is directly proportional to C_D.

Although the general agreement between these results is very encouraging, the overall accuracy is limited by the accuracy of the measurements and the computational model. One example of a modeling approximation is the omission of compressibility effects, which become significant as velocity approaches Mach 1. Further, since the weight of the plane decreases as the engines consume fuel, the thrust required to maintain cruising altitude and velocity continually decreases during the flight.

What If

Budget constraints require smaller engines with less thrust. Thus, less drag can be accommodated. Find an explicit formula that expresses the thrust in terms of the weight, lift coefficient, and other parameters (excluding the drag coefficient). What happens to the thrust requirement as α decreases? What do you think can be done to the airplane design to reduce the constant α in the expression $C_D = C_{D_0} + \alpha C_L^2$?

3-6 FUNCTIONS

The use of substitutions to evaluate an expression for different sets of values is fairly natural, but can become rather tedious. A more mathematical approach is to define a *function* with the appropriate set of one or more arguments and then simply evaluate the function for different combinations of arguments.

Try It

The first step toward defining functions in Maple is to realize that the command **g(x) := x^2**; defines only the name **g(x)**. Verify that **g(x);** returns the expression x^2, but **g(0)**, **g(y)**, and **g(2x)** all return unevaluated.

There are many ways to create functions in Maple. Two of the most common, and powerful, use the arrow (**->**) operator (see **?operators,functional**) and the **unapply** command. In both cases, the resulting Maple function is a rule that determines how a set of input arguments is to be manipulated to form the desired function value (which can be a set, list, expression, numerical value, or any other valid Maple object).

EXAMPLE 3-14

Defining and Evaluating Maple Functions

Create a Maple function equivalent to $f(x) = x^2 + 3x$. Evaluate this function for $x = -2$, $x = 1 + \sqrt{2}$, and $x = \sin(\theta)$.

SOLUTION

The arrow operator definition of this function would appear as

```
> f := x -> x^2 + 3*x;
```

$$f := x \rightarrow x^2 + 3x$$

Note that this creates a rule that takes a single input argument, x, and returns the value $x^2 + 3x$. The three function values requested in this problem are:

```
> f( -2 );
```

$$-2$$

```
> f( 1+sqrt(2) );
```

$$\left(1 + \sqrt{2}\right)^2 + 3 + 3\sqrt{2}$$

which simplifies to

```
> simplify( f( 1+sqrt(2) ) );
```

$$6 + 5\sqrt{2}$$

and

```
> f(sin(theta));
```

$$\sin(\theta)^2 + 3\sin(\theta)$$

If you find yourself calling **simplify** after almost every function call, you should note that the call to **simplify**, or any other Maple function, can be included in the definition of the function, for example,

```
> f := x -> simplify( x^2 + 3*x );
```

. . . .

The **unapply** command is useful for the conversion of an expression into a function. For example, the two roots of the general quadratic polynomial have previously been obtained as a list:

```
> EQN := a*x^2 + b*x + c = 0 :
> ROOTS := [ solve( EQN, x ) ];
```

$$ROOTS := \left[\frac{1}{2}\frac{-b+\sqrt{b^2-4\,a\,c}}{a}, \frac{1}{2}\frac{-b-\sqrt{b^2-4\,a\,c}}{a} \right]$$

You might think that this result could be converted into a function, **R**, with the command

```
> R := (a,b,c) -> ROOTS;
```

$$R := (a, b, c) \text{ -> } ROOTS$$

```
> R(1,1,0);
```

$$\left[\frac{1}{2}\frac{-b+\sqrt{b^2-4\,a\,c}}{a}, \frac{1}{2}\frac{-b-\sqrt{b^2-4\,a\,c}}{a} \right]$$

But, as this example shows, the right-hand side of the arrow operator is not evaluated in the definition of **R**, and there is no connection between the arguments *a*, *b*, and *c* and the names *a*, *b*, *c* in **ROOTS**. The conversion of the list **ROOTS** into a function can be achieved with the use of **unapply**:

```
> R := unapply( ROOTS, (a,b,c) );
```

$$R := (a, b, c) \rightarrow \left[\frac{1}{2}\frac{-b+\sqrt{b^2-4\,a\,c}}{a}, \frac{1}{2}\frac{-b-\sqrt{b^2-4\,a\,c}}{a} \right]$$

```
> R(1,1,0);
```

$$[0, -1]$$

Note the use of the arrow operator notation in the output from the **unapply** command.

EXAMPLE 3-15   ## Symbolic Roots of a Quadratic Function Revisited

Use the function **R** to determine the roots of the general quadratic equation when $c = 0$, when $b = 2$, and when $b = 2$ and $c = 0$.

SOLUTION

Since **R** is already defined, all that remains is to evaluate the function for three different sets of arguments (and to simplify the results).

> R(a,b,0);

$$\left[\frac{1}{2} \frac{-b + \sqrt{b^2}}{a}, \frac{1}{2} \frac{-b - \sqrt{b^2}}{a} \right]$$

> simplify(R(a,2,c));

$$\left[\frac{-1 + \sqrt{1 - a\,c}}{a}, -\frac{1 + \sqrt{1 - a\,c}}{a} \right]$$

> simplify(R(a,2,0));

$$\left[0, -\frac{2}{a} \right]$$

As expected, these results agree with those found in Examples 3-12 and 3-13.

. . . .

Try It   Use the data from Application 3 to create a function that can be used to

obtain the drag for any value of the coefficient of lift.

Maple provides an extensive collection of built-in functions. Each of these functions, as well as any other Maple command, can be used when defining a function. Some of the more common built-in functions include **abs**, **exp**, **ln**, **log**, **log10**, **sqrt**, **frac**, **Re**, **Im**, **max**, **min**, and the trigonometric and inverse trigonometric functions; a full list of initially defined functions can be found in the online help for the topic **inifcns**.

3-7 EXACT VS. APPROXIMATE ARITHMETIC

All Maple computations are performed using as much precision as possible. For example, the constants π and e (= **exp(1)**) are not replaced with their *floating-point approximations* unless explicitly directed to do so with the **evalf** command. The default number of significant digits is stored in the environment variable **Digits**. The number of significant digits used in floating-point operations can be modified by assigning a new value to **Digits** or by specifying a second argument in **evalf**.

EXAMPLE 3-16   ## Exact and Floating-Point Evaluation of a Function

Use the function **R** defined in Example 3-15 to compute the exact roots of $x^2 - 3x - 1$. Find floating-point approximations to these roots using the default number of significant digits and with three significant digits.

SOLUTION

The (exact) roots of this quadratic equation are

> `exact := R(1,-3,-1);`

$$exact := \left[\frac{3}{2} + \frac{1}{2}\sqrt{13}, \frac{3}{2} - \frac{1}{2}\sqrt{13} \right]$$

Floating-point approximations of these roots obtained with the default number of significant digits is

> `default := evalf( exact );`

$$default := [3.302775638, -.302775638]$$

If the calculations are performed using only three significant digits, the approximations are

> `three := evalf( exact, 3 );`

$$three := [3.31, -.31]$$

Observe that the roots computed with three significant digits are not simply the values in default rounded to three decimal places. This is because the previous command computes all quantities in the exact solution using floating-point arithmetic with three significant digits. To obtain the rounded values of the more accurate approximations, use

> `round3 := evalf( default, 3 );`

$$round3 := [3.30, -.303]$$

· ·

Try It   Use **subs** to substitute the values in **exact**, **default**, **three**, and **round3** into $x^2 - 3x - 1$. How many digits of accuracy are obtained with each set of solutions?

Another technique for invoking floating-point arithmetic is to include one or more floating-point numbers in an arithmetic statement. For example, consider the following command in which the two function calls differ only in the inclusion of the decimal point in the right-hand side:

```
> f( sqrt(5) ) = f( sqrt(5.) );
```

$$5 + 3\sqrt{5} = 11.70820394$$

Floating-point computations can also be used to solve equations whose solutions cannot be written in closed form, for example, polynomials of degree five and higher. In these cases, you use the **fsolve** command instead of solve followed by **evalf**.

EXAMPLE 3-17

Approximate Solutions to $y = \tan(x)$

The graphs of $y = \tan(x)$ and $y = x$ intersect an infinite number of times (see Figure 3-1). Find the first five positive points of intersection of these two curves.

Figure 3-1
Graphs of $y = \tan(x)$ and $y = x$

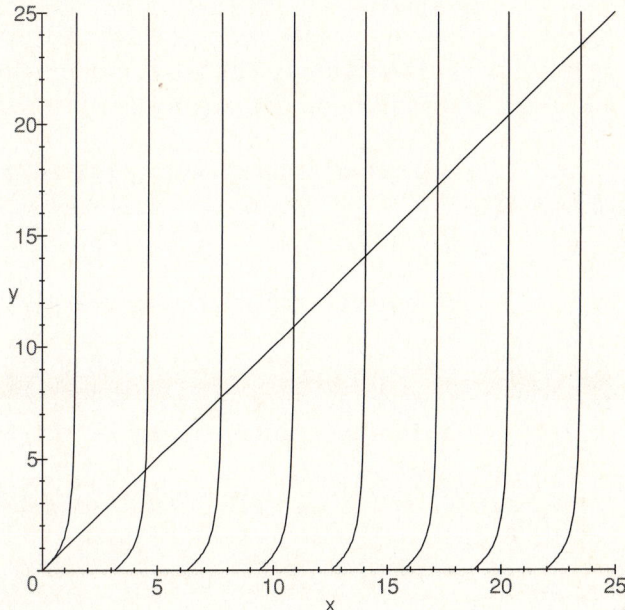

SOLUTION

The two functions are plotted in Figure 3-1. The naive solution is to use **solve**

```
> naive := solve( tan(x)=x, x );
```

$$naive := \text{RootOf}(\tan(\_Z) - \_Z)$$

and then to use **allvalues** to force explicit evaluation of the **RootOf**:

```
> allvalues( naive );
```

$$0$$

This yields only the trivial solution $x = 0$.

The next most obvious approach is to replace **solve** with **fsolve**:

```
> fsolve( tan(x)=x, x );
```

$$4.493409458$$

Based on the information in Figure 3-1, this appears to be the smallest positive intersection of the two curves. To obtain the next four smallest roots, observe that the graphs of $y = x$ and $y = \tan(x)$ intersect exactly once in each period of the tangent function. The first interval to consider is $-\pi/2 \leq x \leq \pi/2$. Here the solution ($x = 0$) is obvious without the use of Maple; moreover, this is not a positive solution. The first positive solution must occur in the next interval of length π:

```
> fsolve( tan(x)=x, x, x=Pi/2..3*Pi/2 );
```

$$4.493409458$$

This is the intersection identified without specifying the optional argument to **fsolve**. The next four intervals of length π should yield the other four solutions that we seek:

```
> fsolve( tan(x)=x, x, x=3*Pi/2..5*Pi/2 );
```

$$7.725251837$$

```
> fsolve( tan(x)=x, x, x=5*Pi/2..7*Pi/2 );
```

$$10.90412166$$

```
> fsolve( tan(x)=x, x, x=7*Pi/2..9*Pi/2 );
```

$$14.06619391$$

```
> fsolve( tan(x)=x, x, x=9*Pi/2..11*Pi/2 );
```

$$17.22075527$$

These roots agree with Figure 3-1.

. . . .

SUMMARY

This chapter introduced some of the fundamental Maple objects—expressions, equations, expression sequences, lists, sets, and functions—and the basic techniques for the manipulation of these objects. Some important lessons were learned about the use of finite-precision arithmetic and floating-point numbers. This chapter's application, obtained from the aeronautical engineering discipline, examined the lift and drag forces on an airplane at cruising altitude.

Key Words

assignment	inequality
assume facility	list
equation	name
expression	operator
expression sequence	set
floating-point approximation	substitution
full evaluation	unevaluated name
function	

Maple Commands

$ (dollar)	log
' (single quotation mark)	log10
-> (arrow operator)	max
:= (assignment)	member
=	min
[ ]	minus
{ }	nops
about	NULL
abs	numer
allvalues	op
assume	Pi
collect	Re
convert	restart
denom	rhs
Digits	RootOf
evaln	seq
exp	simplify
frac	solve
fsolve	sort
Im	sqrt
intersect	subs
lhs	unapply
ln	union

References

1. Anderson, J. D. *Introduction to Flight,* 3rd ed. New York: McGraw-Hill, 1989.
2. McCormick, B.W. *Aerodynamics, Aeronautics, and Flight Mechanics*, 2nd ed. New York: John Wiley & Sons, 1995.
3. Smith, B. *Aerodynamics for Engineers*. 2nd ed. New York: Prentice-Hall, 1989.

Problems

1. (a) Use Maple's **solve** command to solve the system of equations
    $.00001\,u + v = 1, -u + v = 0.$

    (b) Rewrite the system with integer coefficients; find the exact (that is, rational) solution to this system.

2. To verify that the solutions found in Problem 1 are correct, use **subs** to substitute the solutions back into both systems of equations. Further, substitute the rational solution into the original system and the floating-point solution into the integer system.

    Note that some numbers are integers and others are floating-point values. There is a difference.

    To illustrate, use the **evalb** command (see **?evalb**) to see if Maple thinks the equations are satisfied. Explain the results.

3. This problem illustrates some of the difficulties that can occur when subtracting floating-point numbers.

    Compute the floating-point approximations of $N1 = 8721\sqrt{3}$, $N2 = 10681\sqrt{2}$, $SUM = 8721\sqrt{3} + 10681\sqrt{2}$, and $DIFF = 8721\sqrt{3} - 10681\sqrt{2}$ using 2, 3, 4, ... , 19, 20 significant digits. To how many digits do **N1** and **N2** agree?

    What are the correct values of **SUM** and **DIFF** to five significant digits? How many floating-point digits are needed to compute **SUM** and **DIFF** to this accuracy?

    A more reliable way to compute the difference is to note that **PROD = expand(DIFF*SUM)** is an integer when fully simplified. (Why?) Thus, **DIFF = PROD/SUM** which can be computed without any subtraction. How many floating-point digits are needed to obtain five significant digits of accuracy in the value of **DIFF** when it is computed by division?

    One moral of this exercise is that the accuracy of a floating-point calculation may not be the same as the number of significant digits used in a calculation. This is a general property of floating-point arithmetic, not just Maple.

4. Use **subs** to verify that both solutions found in Example 3-11 are, in fact, points of intersection of the two curves. In general, there are two solutions. Find values of r for which there are no solutions and a single solution. Can there ever be three points of intersection?

5. Calculate the speed of sound in air at sea level and at 35,000 feet (in m/s and in ft/sec) using the formulas provided in the text.

6. It is clear that the weight of an airplane decreases as fuel is consumed. Therefore, the lift required to maintain the cruising altitude will decrease as fuel is consumed. This has not been taken into account in the application. Given a particular fuel consumption rate and starting weight, the distance, s, a plane can travel is given by

$$s = \frac{V \ln\left(\dfrac{m_0}{m}\right)\left[\dfrac{L}{D}\right]}{TSFC\, g}$$

where *TSFC* is the Thrust-Specific Fuel Consumption, g is the gravitational acceleration (32.1740 ft/sec^2), m_0 is the initial mass, and m is the final mass. Assuming that $TSFC = 0.75$ lb$_m$/lb$_f$-hr and that the maximum fuel capacity is 180,000 lb, determine the maximum range based on the lift and drag results required for level flight at 35,000 feet. Determine the minimum amount of fuel required for this aircraft to fly across the United States (approximate distance of 3500 miles).

(The preceding formula was derived by Breguet. The derivation of this equation, which involves differential equations, will be explored in more detail in Problem 13 in Chapter 6.)

(*Hint:* $\dfrac{m_0}{m} = 1.56$. Watch the units; *TSFC* has hours, not seconds.)

7. Express the thrust needed to keep an aircraft at cruising altitude in terms of the aircraft's weight, aspect ratio, wing span, and Mach number when altitude is 35,000 feet and the lift-to-drag coefficients are $(C_{D_0}, \alpha) = (0.0155, 0.0588)$. As an aeronautical engineer, explain what changes in the aircraft's weight, wing span, aspect ratio, and Mach number would decrease the thrust requirement.

8. Determine the range of an airplane at cruising altitude in terms of its "empty" weight (that is, no passengers and no fuel), and in terms of the amount of fuel, wing span, Mach number, aspect ratio, *TSFC*, α, γ, δ, and σ. Give your answer in miles. Determine whether cruising 3,000 feet above or below the 35,000 foot cruising altitude increases the aircraft's range. Use $\delta_{32000} = 0.2707$, $\delta_{38000} = 0.2037$, $\sigma_{32000} = 0.3471$, and $\sigma_{38000} = 0.2692$.

9. Fuzzy sets are used in an increasing number of engineering disciplines to more accurately mimic the manner in which human beings make decisions. This general area of study is often referred to as fuzzy logic. For example, a fuzzy logic decision-making circuitry could be incorporated into the timing mechanism of a dishwasher to determine to what extent the dishes within are clean or dirty or, more important, partially dirty. In this fashion, the dishwasher could be made to operate more efficiently if fuzzy logic can be used to shut off the dishwasher as soon as the dishes are determined to be clean instead of simply running for a fixed amount of time.

An example of a fuzzy set is "numbers close to 10." A traditional "crisp" set would, for example, give a value of 1 to all numbers between 8 and 12 to indicate full membership in this set. All numbers outside this range would get a value of 0 to indicate nonmembership. But this is not realistic since the number 7.9 should have greater membership status than the number 2.5. Suppose, instead, we use a fuzzy set to describe numbers close to 10 by using the membership function

$$\mu(x) = \frac{1}{1 + (x - 10)^2}.$$

Note that $x = 10$ has full membership status, since $\mu(10) = 1$, and all other values of x have partial membership status, with values closer to 10 having a status closer to full membership. In fuzzy logic, non-membership in a fuzzy set A is determined by the membership function for $C = A$, the complement of A; that is, if μ_A is the membership function for A, then the membership function for C is $\mu_C = 1 - \mu_A$. (Note that when $A = \{10\}$, $\mu_C(10) = 1 - \mu_A(10) = 0$ so that the number 10 has nonmembership in C while all other values of x have partial membership in C.)

(a) Define and plot the membership functions for $A = \{10\}$ and the complement of A.

(b) Define and plot the membership function for $B = \{15\}$.

(c) Determine a membership function for the union of A and B, that is, for the numbers close to 10 OR close to 15.

   (*Hints:* What properties should this function have? Plot the membership function for A and for B on the same axis. How can these functions be combined to create a function with the necessary properties? See **?max**.)

(d) Determine a membership function for the intersection of A and B, that is, for the numbers close to 10 AND close to 15.

   (*Hints:* This membership function never takes on the value 1 since no element has full membership in both A and B.)

10. This is a continuation of the Try It! exercise first discussed at the end of Section 3-2.

The exact division points for the 128 levels are not convenient for human analysis; the floating-point form of these levels are almost equally difficult to use. What is needed is a list of the levels with only a few decimal places. Create the list of levels with exactly two decimal digits of accuracy.

4 Plotting and Analyzing Engineering Functions and Data

Bandwidth of an Optical Filter

A filter is a device that allows only a certain class of properties to pass and blocks others. Optical engineers frequently analyze filters that limit the band of frequencies (or colors in the visible range) that are allowed to pass through an optical system. Simple examples of optical filters include sunglasses, which are often designed to block ultraviolet frequencies while passing the visible portion of the spectrum in a somewhat attenuated fashion, and the special lenses in traffic signals that transmit only red, yellow, or green light. Electrical engineers often use filters in a wide variety of applications, one example being the design of a radio receiver.

Two fundamental characteristics of a certain class of optical filters called interference filters (or Fabry–Perot etalons) are bandwidth and finesse. The application in this chapter examines these characteristics. You will use both the graphical features available in Maple and the symbolic and numeric features introduced previously to better understand an interference filter's transmission properties.

INTRODUCTION

T he graphical display of information, whether in the form of an explicit expression, a list of data, or the solution of an equation, can be a very powerful engineering tool. In this chapter, you will learn how to use Maple to create a variety of plots, including two- and three-dimensional plots of one or more functions or sets of tabulated data. You will see how a sequence of plots can be combined to form an animation. You will also learn how to increase the effectiveness of a plot by the appropriate use of titles, labels, colors, and other information.

4-1 PLOTTING FUNCTIONS AND EXPRESSIONS

Maple's **plot** command is capable of graphing a variety of objects, including functions and expressions. The basic ingredients required by the **plot** command are (1) the expression to be plotted, (2) the independent variable, and (3) the range of values for the independent variable (the domain). Optional arguments can be used to specify a title, axes labels, viewing rectangle, or other features that improve the effectiveness of the plot. The online help worksheets for **plot** are easily accessed by a topic search for **plot**; the complete list of optional arguments is listed under the topic **plot,options**. The corresponding information for three-dimensional plots can be found using the keywords **plot3d** and **plot3d,options**.

The **plots** package contains additional plotting commands, including **display**, **animate**, and **implicitplot**. The **with** command is used to load a Maple package. Here, the command **with(plots);** is used to load the plots package.

```
> with( plots );
```

[*animate, animate3d, changecoords, complexplot, complexplot3d,
conformal, contourplot, contourplot3d, coordplot, coordplot3d,
cylinderplot, densityplot, display, display3d, fieldplot, fieldplot3d,
gradplot, gradplot3d, implicitplot, implicitplot3d, inequal, listcontplot,
listcontplot3d, listdensityplot, listplot, listplot3d, loglogplot, logplot,
matrixplot, odeplot, pareto, pointplot, pointplot3d, polarplot,
polygonplot, polygonplot3d, polyhedraplot, replot, rootlocus,
semilogplot, setoptions, setoptions3d, spacecurve, sparsematrixplot,
sphereplot, surfdata, textplot, textplot3d, tubeplot*]

The **display**, **animate**, and **implicitplot** commands can be found in the list of commands defined in the **plots** package. More specific information about the contents of this package will be given later in this chapter. Full information about the **plots** package is available in the online help (see **?plots**).

Plotting One Function

The basic syntax for the **plot** command is **plot(*expr*, *var* = *a* .. *b*);** where ***expr*** is a Maple expression with no more than one parameter, ***var*** is the name of the independent variable, and ***a* .. *b*** is the interval on which the first argument should be plotted.

EXAMPLE 4-1

Plotting a Periodic Function

Plot two periods of $v = 2 + \frac{1}{2} \sin(\theta) \cos(2\,\theta)$.

SOLUTION

```
> restart;
```

The period of $\sin(\theta)$ is 2π, and the period of $\cos(2\theta)$ is π, so the product has period 2π. The requested plot will be displayed over any interval of length 4π, for example, $-2\pi \leq \theta \leq 2\pi$. A Maple plot of this function can be obtained with the command

```
> v := 2 + sin(theta) * cos(2*theta)/2;
```

$$v := 2 + \frac{1}{2} \sin(\theta) \cos(2\,\theta)$$

```
> plot( v, theta = -2*Pi .. 2*Pi );
```

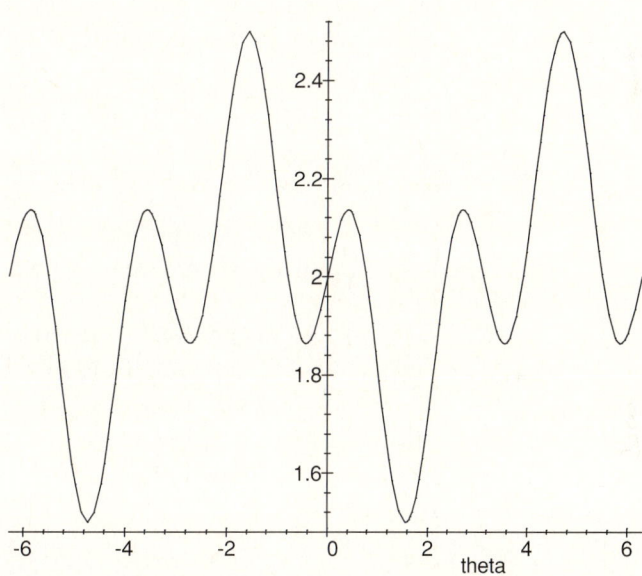

Note that the period of v is, indeed, 2π. Notice, also, how Maple automatically selects the vertical scale to show as much detail in the plot as possible. The vertical range and vertical label can be specified as optional arguments of **plot**.

. . . .

Try It Replot the function v described in Example 4-1 with a vertical range that begins at 0. (*Hint:* Consult the online help for **plot** for the necessary optional arguments, if needed.)

Plotting Multiple Functions in One Plot

Sometimes you want to display the graphs of two or more functions in the same plot. When the first argument of **plot** is a list or set of expressions or functions, Maple creates a single plot containing the graph of each element in the first argument. In these situations it is often useful to use the optional argument **color=** , where the right-hand side contains a list of colors to be used for the different plots (see **?plot,colors**). Likewise, the **style=** optional argument can be used to choose between plotting the individual points in a graph (**style=POINT**) or connecting the points with straight lines (**style=LINE**).

EXAMPLE 4-2

Two Graphs in a Single Plot

Display the graphs of both $v_1 = \sin(\theta)$ and $v_2 = \frac{1}{2}\cos(2\theta)$ in a single plot.

SOLUTION

To illustrate the plotting of both expressions and functions, define

```
> v1 := sin( theta ) ;
> v2 := theta -> cos(2*theta) / 2 ;
```

$$v_1 = \sin(\theta)$$
$$v_2 = \frac{\cos(2\,\theta)}{2}\ .$$

Then the plot of the expression v_1 and the function v_2 can be obtained using

```
> plot( [ v1, v2(theta) ], theta = -2*Pi .. 2*Pi,
>       color=[red,blue], style=[point,line] );
```

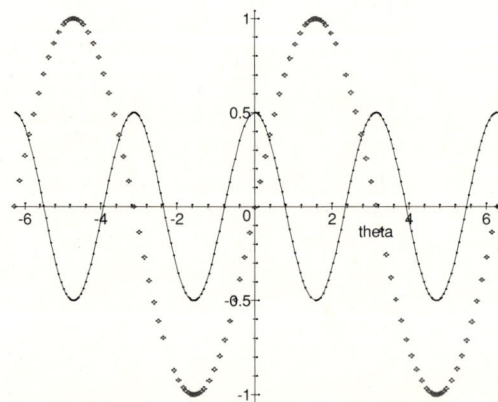

Note that since a printed version of the plot may be black and white (or grayscale), colors should not be used as the sole means of identifying curves in a plot; distinct styles serve nicely to distinguish functions in a plot.

It is a good habit to consider how a plot will be printed or displayed when customizing the appearance of a plot. Another good habit to develop is to title all plots; this is done with the **title=** option. Text labels can also be used to identify specific features within a plot.

It is also possible to plot a set of functions. However, since the elements of a set are not ordered, it is not possible to predict the color or style used for a particular function in the set.

. . . .

Try It Repeat Example 4-2 with the first argument specified as a set of functions.

Combining Multiple Plots into a Single Plot

There are many situations in which it is desirable to combine several Maple plots into one composite plot. The previous section described how this can be done using the **plot** command when the same options are to be used for each of the plots. When different options (other than **color=** and **style=**) are desired for individual plots, the **display** function from the **plots** package should be used.

EXAMPLE 4-3

The display Command

Create a single plot containing the graph of two periods of v and a single period each of v_1 and v_2. Be sure to choose options that enable each function to be clearly identified.

SOLUTION

If all three functions are to be plotted on the same domain, this could be completed with a single **plot** command. However, the instructions specifically require that the domains for v, v_1, and v_2 be of length 4π, 2π, and π, respectively. To combine the three plots into a single plot requires the **display** command from the **plots** package. The first step is to load the plots package:

```
> with( plots ):
```

Next, create—but do not display—the three plots that will be needed to form the final image.

```
> P1 := plot( v, theta = -2*Pi .. 2*Pi ):
> P2 := plot( v1, theta = 0 .. 2*Pi, color=GREEN, style=POINT ):
> P3 := plot( v2(theta), theta = -Pi .. 0, color=BLUE, linestyle=2 ):
```

Note that none of these plots is displayed when it is created; rather, each plot is assigned a name for future reference. If a semicolon is used with a **plot** command whose result is assigned to a Maple name, the complete Maple plot structure is displayed. This data structure is not something the general user needs or wants to see. For this reason, you are strongly encouraged to terminate the **plot** command with a colon in situations where the plot is assigned a name.

Now, back to the problem of superimposing the plots **P1**, **P2**, and **P3** onto one set of axes. You will use the **display** command to specify options for the composite plot.

```
> display( { P1, P2, P3 },
>           title='Three Graphs Combined into One Plot' );
```

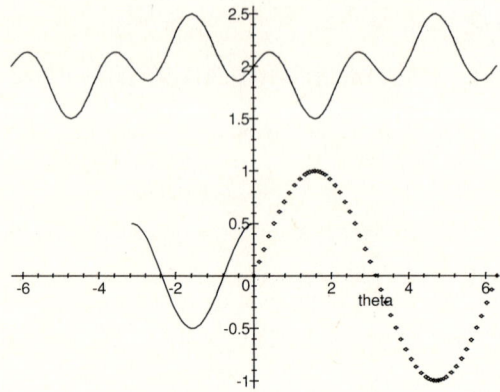

Three Graphs Combined into One Plot

Note that options concerning the individual functions to be plotted (for example, the domain, range, **color=**, **style=**, **linestyle=**) are specified in the individual **plot** commands; options that refer to the composite plot (for example, **title=**, **axes=**) are specified in the **display** command. In certain situations, it may be necessary to experiment with these options to achieve the desired final image.

Animation

In some situations, it is more appropriate to display a set of plots as an ordered sequence of frames—an *animation*—rather than as a single composite plot of several functions. The **animate** command, also from the **plots** package, is used for this. The basic syntax is the same as for **plot**, except that the expression to be plotted should contain two independent variables (one to be used as the independent variable in the plot, the other to be used as the animation parameter); furthermore, the optional argument **frames=** can be used to control the number of frames in the animation (see **?animate**).

EXAMPLE 4-4   ## Animating a Family of Functions

Produce a ten-frame animation of the family of functions

$$y = 2 + \frac{\sin(x)\cos(ax)}{a} \quad \text{where } 1 < a < 10.$$

SOLUTION

The **animate** command that produces this animation is very simple. However, as a increases, the default resolution of the frames is not acceptable; the **numpoints=** optional argument (see **?plots,options**) specifies the number of points to be evaluated for each image.

```
> animate( 2 + 1/a * sin(x) * cos(a*x), x = -2*Pi .. 2*Pi,
>            a = 1 .. 10, frames = 10, numpoints=200 );
```

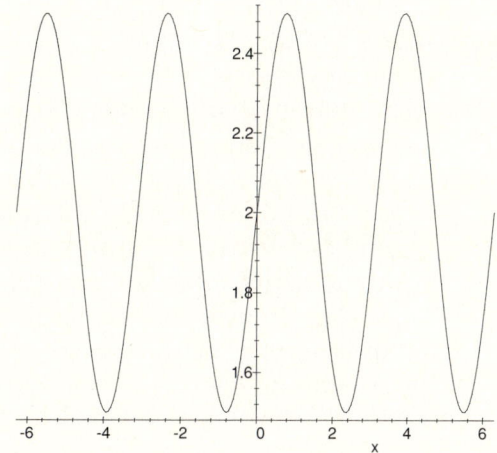

As the previous plot shows, only the first frame is initially displayed. Subsequent frames in the animation can be viewed either by using the Animate menu on the menu bar or the Animate context bar on the control bar (see Figure 4-1). These controls are accessible only when the current context is an animation region. The Animate menu bar contains a set of VCR-like controls: the solid box is the Stop button, the solid arrow is the Play button, the small arrow with a vertical bar icon is the Frame Advance button, the single arrows specify the direction of the animation, the double arrows increase and decrease the frame rate, which is displayed on the left end of the status line. The two rightmost icons determine whether the animation is played in single-cycle or continuous-cycle mode. (The balloon help can be useful until you learn these icons.)

Figure 4-1
Menu, tool, and context bars for animations

Menu bar

Tool bar

Context bar

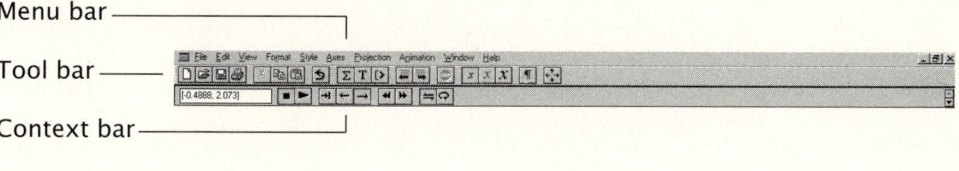

If frames corresponding to specific parameters are desired, it is recommended that you create the individual plots, and then use **display** with the **insequence=true** option to create the animation.

Try It Use **display** and the **insequence=true** option to create an animation that shows the graphs of v, v_1, and v_2 as a sequence of three frames.

Customized Plotting Features

Some fine-tuning of a plot can be done through the Maple interface. To make the plot active, click the left mouse button in the plot region. When the graphics (or animation) region is active, an anchored outline displays the boundary of the plot, and the Plot (or Animation) context bar will appear on the control bar. A variety of plot options—including the thickness and style for line plots and symbol and size for point plots, the location of axes, and the aspect ratio—can be altered using the icons on the Plot context bar; the same controls are also available on the Style, Axes, and Projection pull-down menus in the menu bar or the pop-up menu that appears when the right mouse button is pressed when the cursor is located in the plot region. The **plots,options** help worksheet contains a complete listing of options for all two-dimensional plot commands. Although the interface controls can be convenient, if the preferred settings are known in advance, it is often easier to use the optional arguments. In fact, when the same options are to be used for several plots, the **setoptions** command (see **?plots,setoptions**) should be used.

EXAMPLE 4-5

Plotting a Discontinuous Function

Create a plot of several periods of $y = \tan x$. Be sure the plot contains only the graph of the tangent function and shows all interesting features of the function. Explain any nonperiodic behavior that you see in the plot. Refine the optional arguments until the plot clearly shows that the function is periodic and discontinuous.

SOLUTION

You might think that this plot could be easily produced using

```
> plot( tan(x), x=-Pi..2*Pi, title='The Tangent Function ?' );
```

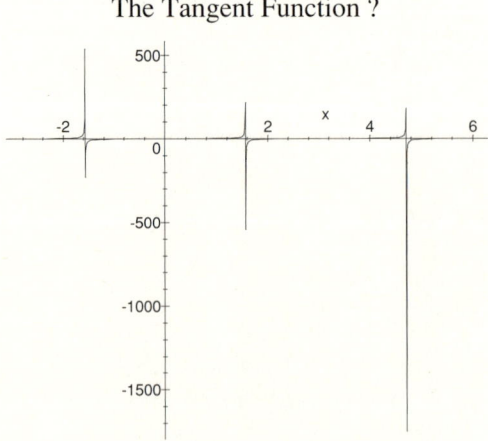

This graph does not have the periodic structure expected of the tangent function. Look carefully at the plot. In particular, look at the labels on the vertical axes. The vertical scale is marked in units of 100, and the vertical range extends from –1700 to 500. On this scale, the details of the plot are not visible.

To understand the steps needed to produce a reasonable plot of the tangent function, it is first necessary to understand a little about how plot selects the points to be plotted. The selection of points from the domain is adaptive—that is, Maple selects more points in regions where the function values exhibit the greatest variation. In the case of the tangent function, this means that some points are selected close to each asymptote. (This can be observed in the above plot by selecting Point from the Style menu that appears in the menu bar when the current focus is in the plot region. The same effect can be obtained with the use of the optional argument **style=POINT**. Selecting **axes=BOXED** further clarifies this point.) One of the best ways to focus on the behavior away from the asymptotes is to restrict the vertical range:

```
> plot( tan(x), x=-Pi .. 2*Pi, y = -15 .. 15,
>        title = 'Almost the Tangent Function' );
```

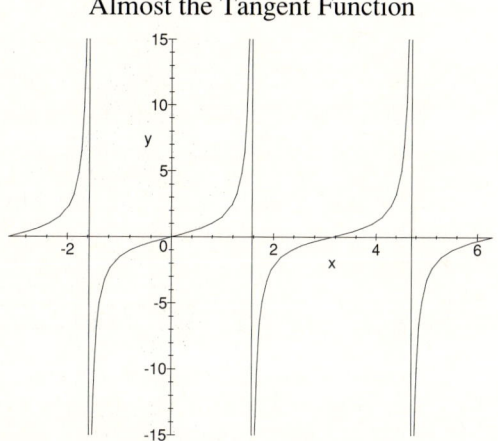

Almost the Tangent Function

Even this plot is not completely acceptable. The (essentially) vertical lines that appear at odd multiples of $\pi/2$ are the *vertical asymptotes*. These are included in this plot since Maple connects the last point to the left of the discontinuity with the first point to the right of the discontinuity. These spurious lines can be eliminated by using the **discont=true** option.

```
> plot( tan(x), x=-Pi..2*Pi, y=-15..15, discont=true,
>      title='The Tangent Function' );
```

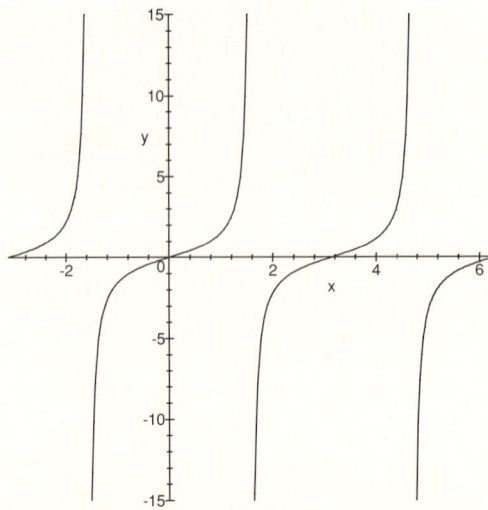

The Tangent Function

Another factor to consider when analyzing graphical data is the *aspect ratio* of the plot; that is, how are the horizontal and vertical scales related? By default, Maple chooses the scaling so that the plot fills the plot region, which is often a square. If the same horizontal and vertical scales are needed, for example, the aspect ratio should be 1, then either the **scaling=CONSTRAINED** optional argument should be specified or the 1:1 icon in the Plot context bar should be clicked.

- -

Application 4   BANDWIDTH OF AN OPTICAL FILTER

Optical engineers are frequently asked to create devices that limit the frequencies (or wavelengths) transmitted through an optical system. In the visible portion of the electromagnetic spectrum, these frequencies correspond to the different colors of light. An optical filter is frequently characterized by its *bandwidth*, which indicates whether a narrow or broad spectrum of colors is transmitted. A clear lens in a camera has a very large bandwidth since it allows virtually all visible colors to pass through it; a red lens has a much narrower bandwidth since it blocks all colors except red. An alternative way to characterize the bandwidth of certain filters called *interference filters* is through the *finesse* of the filter, which is the ratio of the separation between transmission peaks to the width of a particular transmission mode. In this application the five-step problem-solving process will be used to discover some of the relationships between the bandwidth and finesse for an optical filter. In particular, an optical interference filter with finesse 20 ± 10% will be designed.

Fundamentals

An *optical interference filter* consists of two partially transmitting parallel mirrors, each having *reflectivity R* ($0 < R < 1$), separated by an air gap of length *L* (see Figure 4-2). When *R* approaches 1, more of the incident ▼

light on a particular mirror is reflected and less is transmitted through that mirror; when R approaches zero, transmission through a particular mirror approaches 100 percent (reflection approaches 0 percent). Figure 4-2 shows a schematic of a beam of light incident on the optical filter, with reflected and transmitted beams resulting from interaction with the filter. A filter with a large bandwidth transmits the light intensity over a wide range of frequencies surrounding a transmission peak. As R approaches 0, this range of frequencies (the bandwidth) increases. On the other hand, if the bandwidth is narrow, the transmitted frequencies occur over a narrower range near the transmission maximum, making the filter more spectrally discriminating. This occurs as R approaches unity.

Figure 4-2
**Optical
interference filter**

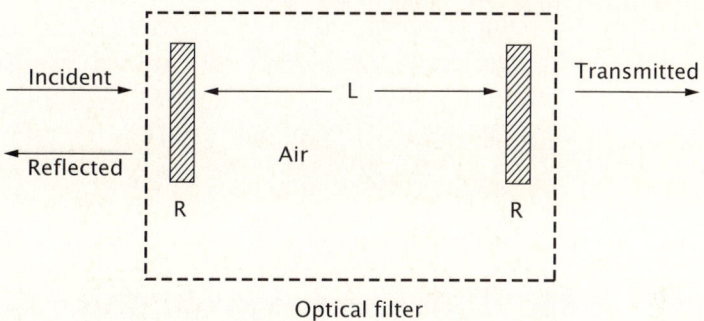

Optical filter

An optical interference filter transmits many different frequencies, but the width of each band of transmitted frequencies is generally the same (see Figure 4-3). The *separation* (Δv) between the frequencies at which the filter transmission peaks occur is inversely related to the mirror separation distance L, so, for example, the frequency separation between the peaks is reduced by a factor of two when the separation distance between the two mirrors is doubled.

Figure 4-3
**Transmitted light
versus frequency**

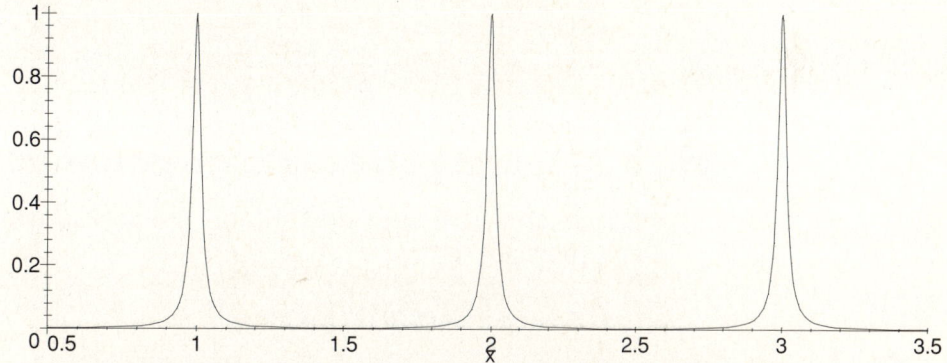

Observe that very little light is transmitted for most frequencies, but all of the light is transmitted at each peak frequency. The bandwidth of each transmission mode is defined to be the *full width at half maximum* (FWHM) bandwidth: that is, the width of frequencies for which exactly half of the incident light is transmitted. The narrower the bandwidth, the more frequency-selective is the optical filter. If $x = 1$ corresponds to a wavelength of 0.6 nanometers (red light), Figure 4-3 represents a red filter with a fairly narrow bandwidth. If the entire curve is shifted so that a peak occurs at a wavelength of 0.48 nanometers (corresponding to a value of x larger than $x = 1$, since wavelength and frequency are inversely related), the filter would transmit only blue light. If the bandwidth increased significantly, the filter would transmit more colors simultaneously; a clear lens that transmits all colors has a relatively large bandwidth.

The finesse, F, of an optical filter is the ratio of the peak separation between transmission modes (Δv) to the frequency width (that is, the bandwidth) of one mode in the transmitted signal. The peak separation is inversely proportional to the length L. Thus, the finesse increases as the bandwidth decreases and separation is not changed. The filter is more frequency-selective when the finesse is large. A filter with a very low finesse does not clearly discriminate between incident frequencies and has a very broad bandwidth. Based on the previous discussion, a large bandwidth should correspond to a filter with mirrors with low reflectivity. The finesse can also be approximated directly in terms of the reflectivity of the mirrors, $F = \dfrac{\pi \sqrt{R}}{1 - R}$; this approximation improves as the finesse increases.

1. Define the problem

You have been given the task of designing an optical interference filter with finesse $F = 20$ with a 10 percent tolerance.

2. Gather information

The only information at this point is the relationship between reflectivity (R) and finesse (F): $F = \dfrac{\pi \sqrt{R}}{1 - R}$.

3. Generate and evaluate potential solutions

A first objective is to understand the relationship between reflectivity and finesse.

```
> F := Pi * sqrt(R)/(1-R);
```

$$F := \frac{\pi \sqrt{R}}{1 - R}$$

Note that the finesse is defined for all $0 \le R < 1$, is small for mirrors with low reflectivity, and has a vertical asymptote at $R = 1$. (We will consider the above equation to be valid for the purpose of our discussion, but recall that the finesse, defined as the ratio of the peak separation to the bandwidth, begins to break down as the reflectivity decreases). More information can be obtained from a plot of F vs. R:

```
> plot( F, R=0..1, 'F'=0..100, title='Finesse vs. Reflectivity' );
```

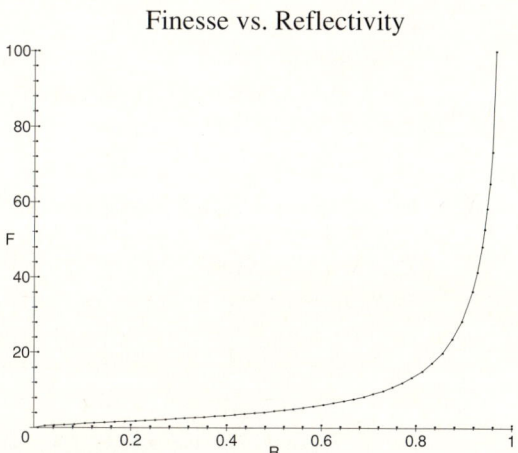

Because F is unbounded near $R = 1$, you are advised to explicitly restrict the vertical viewing range. Here the third argument restricts the vertical scale to the interval ($0 \le F \le 100$) and also specifies a label for the vertical axis. The single quotation marks (**'**) prevent Maple from replacing the name **F** with its current value (which would be an invalid argument—to understand this, try executing the previous **plot** command without the single quotation marks).

The plot is consistent with the discussion about the relationship between finesse and reflectivity. In particular, low reflectivity corresponds to poor frequency discrimination, and therefore low finesse and high reflectivity corresponds to highly selective frequency transmission and very high finesse.

A first approximation to the reflectivity that corresponds to $F = 20$ is obtained by clicking the left mouse button on a point on the curve with F close to 20. For example, (0.8578, 20.26) appears to be very close to the curve. A more accurate approximation can be obtained by zooming in on the plot:

```
> plot( F, R=0.8..0.9, title='Finesse vs. Reflectivity -- ZOOM' );
```

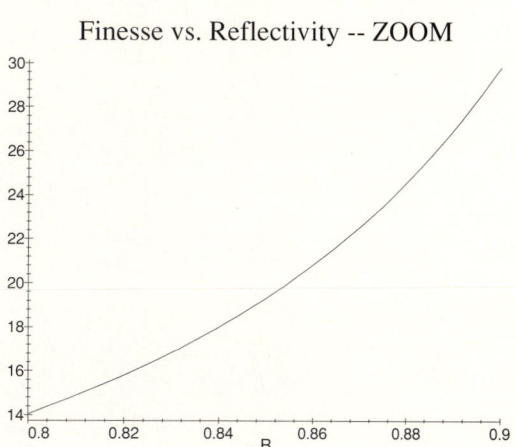

On this scale, the point (0.8547, 20.02) appears to be very close to the curve. These estimates suggest that, to three significant digits, a pair of mirrors that reflect 85.5% of the incident light should have a finesse very close to 20.

Note that these approximations depend on a number of local factors; your estimates may be somewhat different from what is reported here.

4. Refine and implement a solution

One way in which an engineer needs to understand the above estimate is by investigating its sensitivity to the precise value of the finesse. This is particularly important in this case since the results of Step 3 are approximate, with an unknown accuracy. Even if the exact reflectivity were known, the manufacturing and installation processes will introduce some variations into the filter. The design criteria for this problem state that these fluctuations need to be controlled so that the finesse is within 10 percent of the desired value: that is, how much can R vary (around $R = 0.855$) and still ensure that the finesse will satisfy $18 \leq F \leq 22$?

Since the finesse is an increasing function of reflectivity, it suffices to estimate the reflectivities that correspond to $F = 18$ and $F = 22$. This information can be obtained from the last plot, but a more sophisticated (and more informative) approach is to include, in addition to the F vs. R plot, the graphs of constant functions at the desired levels of the finesse.

```
> plot( [F,18,20,22], R=0.8..0.9, color=[BLACK,BLUE,RED,BLUE],
>         title='Finesse vs. Reflectivity ( F=20 +- 10% )' );
```

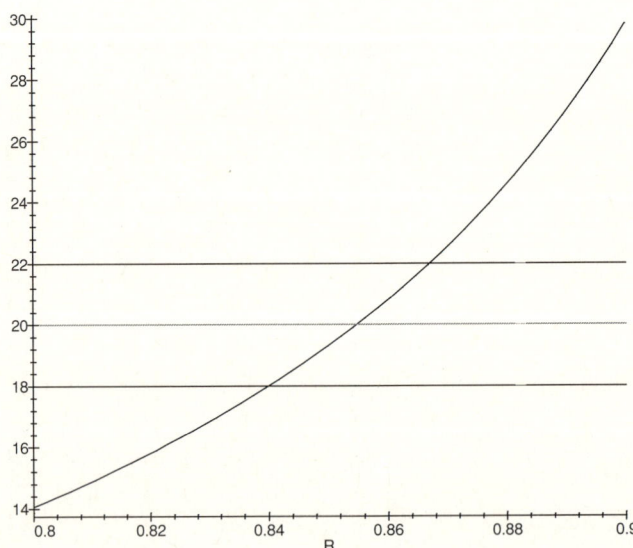

Finesse vs. Reflectivity (F=20 +- 10%)

This plot can be used to find that the finesse is approximately 18 when $R = 0.840$ and approximately 22 when $R = 0.867$. Thus, the reflectivity of the mirror needs to satisfy $0.840 \leq R \leq 0.867$.

```
> abs(0.867-0.855)/0.855, abs(0.840-0.855)/0.855;
```

$$.01403508772, .01754385965$$

That is, the design specifications of $F = 20$ can be maintained to within 10 percent only if the manufacturing tolerance for the mirror reflectivities can be held to within about 1.4%.

5. Verify and test the solution

The estimates found in Steps 3 and 4 depend on the resolution of the plot and the precise placement of the mouse in the plot window. An analytic solution can be found (by hand), and will be more accurate. First, the reflectivity that corresponds to a finesse of $F = 20$ is found by solving the equation $F = 20$ for R:

```
> R20 := solve( F=20, R );
```

$$R20 := -\frac{1}{20}\pi\left(-\frac{1}{40}\pi + \frac{1}{40}\sqrt{\pi^2 + 1600}\right) + 1$$

or, as a floating point number,

```
> R20 := evalf( R20 );
```

$$R20 := .8547736445$$

This value, to three significant digits, is remarkably close to the value obtained directly from the graph. In the same manner, the sensitivity analysis yields

```
> R18 := evalf( solve( F=18, R ) );
> R22 := evalf( solve( F=22, R ) );
```

$$R18 := .8400346341$$
$$R22 := .8670326771$$

which correspond to relative errors of

```
> abs( R22 - R20 )/R20, abs( R18 - R20 )/R20;
```

$$.01434184673, .01724317367$$

These results show considerable agreement with the original estimates obtained from the graph. To truly verify and test these solutions, you need to fabricate the optical filter within the design tolerance to see whether it does indeed produce a measured result having a finesse between 18 and 22. It turns out that the actual finesse will also depend on mirror flatness and the angular spread of the light beam.

What If

What if the project constraints call for a more discriminating optical filter than the one in which $F = 20$? This means that the filter must have a narrower bandwidth. What does this imply regarding the band of frequencies that are transmitted? Would the finesse of the new filter be larger or smaller than $F = 20$? What is the relative change in the reflectivity corresponding to a doubling of the finesse?

4-2 MORE PLOTTING COMMANDS

The **plot**, **display**, and **animate** commands are probably the most frequently used Maple plotting commands. In this section you will see additional plotting commands that can be used with implicitly defined functions, parametric curves, and vector fields. All of these commands are found in the **plots** package.

Implicit Functions

Some functions are not easily expressed in the (explicit) form $y = f(x)$, but do have a convenient definition in the implicit form $F(x, y(x)) = 0$. For example, the points on the unit circle centered at the origin can be expressed either in terms of two explicitly defined functions $y = \sqrt{1 - x^2}$

and $y = -\sqrt{1 - x^2}$ or in terms of the single implicit equation $x^2 + y^2 = 1$. A plot of the curve defined by the points that satisfy a function of two variables can be obtained using the **implicitplot** command from the **plots** package. The basic arguments to **implicitplot** are the equation that defines the points on the curve and a range for each of the two variables in the equation; optional arguments are the same as in the standard **plot** command. Thus, a plot of the unit circle can be obtained with

```
>  implicitplot( x^2 + y^2 = 1, x = -1 .. 1, y = -1 .. 1,
>          title='Unit Circle (implicit)',
>          scaling=CONSTRAINED );
```

Unit Circle (implicit)

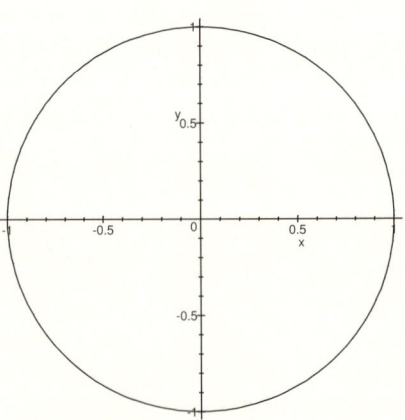

| EXAMPLE 4-6 | **Plotting an Ellipse** |

Plot all points on the ellipse with horizontal semi-axis 4 and vertical semi-axis 2 which is centered at the point (1, –2).

SOLUTION

The implicit equation that describes this ellipse is

> ELLIPSE := (x-1)^2/4^2 + (y+2)^2/2^2 = 1;

$$ELLIPSE := \frac{1}{16}(x-1)^2 + \frac{1}{4}(y+2)^2 = 1$$

A little thought shows that the x-coordinates of these points fall in the range $|x-1| \leq 4$. Similarly, the range of values of the y-coordinates is $|y+2| \leq 2$. Then

> implicitplot(ELLIPSE, = -3 .. 5, y = -4 .. 0,
> title='Ellipse: center (1,-2), axes: 4 (hor) and 2 (ver)');

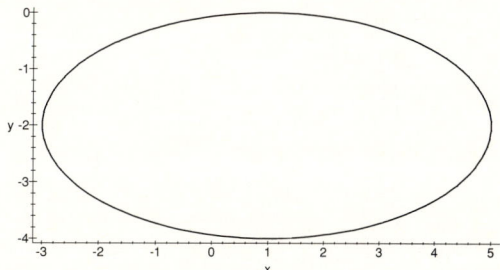

Ellipse: center (1,-2), axes: 4 (hor) and 2 (ver)

Note that this curve may appear as a circle when Maple chooses the aspect ratio; to see this as an ellipse, set the aspect ratio to 1 as described previously. Resizing the plot window can also change the aspect ratio. If the window is made too short, the title or axes labels can interfere with the plot. Try to re-create the plot exactly as it appears in the text.

· ·

Try It

Modify the plot of the ellipse to include the major and minor axes as dashed lines with different colors. (*Hint:* You may wish to use **implicitplot** to draw one or both of the line segments.)

Try It

To solve this problem, refer to Application 4. The transmission function, T, gives the ratio of light that passes through a filter to that which enters the filter. When x is proportional to the frequency of the light and F is the finesse, the transmission function is $T = \dfrac{1}{1 + \left(\dfrac{2\,F\sin(\pi x)}{\pi}\right)^2}$. Use

implicitplot to plot the points (x, F) where $T(x, F) = \dfrac{1}{2}$. Use this plot to compute the FWHM for $F = 20$ and to confirm that a higher finesse corresponds to more discriminating filters.

Parametric Curves

Another large class of curves can be defined in parametric form $(x(t), y(t))$, $a \le t \le b$, where x and y are continuous functions. For example, the unit circle can be written in the parametric form $(\cos(\theta), \sin(\theta))$, $0 \le \theta \le 2\pi$.

The standard **plot** command is used to plot *parametric curves* in Maple. The first argument of **plot** should be a three-element list; the first two arguments are the expressions for the two coordinates of each point on the curve, and the third element of the list is an equation that identifies the parameter of the curve and the range of values for the parameter. For example, a parametric plot of the unit circle can be obtained using

```
> plot( [ cos(theta), sin(theta), theta=0..2*Pi ],
>        title='Unit Circle (parametric)', scaling =CONSTRAINED );
```

Unit Circle (parametric)

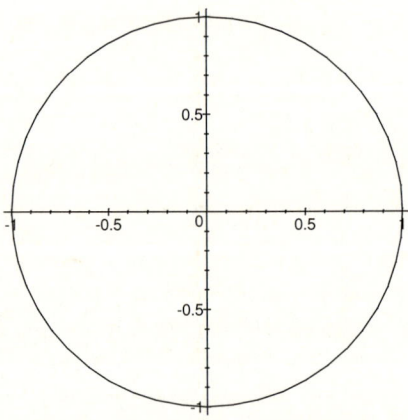

EXAMPLE 4-7   An Implicit Plot of an Algebraic Curve

Plot all points (x, y) with $|y| < 1$ that satisfy $x^2 = y^2(1 - y)$.

SOLUTION

This appears to be a problem that should be solved using **implicitplot**. The constraint that $|y| < 1$ imposes a similar constraint on x: $|x|^2 = |y|^2 |1 - y| \leq 2$, so that $|x| \leq \sqrt{2}$.

```
> implicitplot( x^2 = y^2 * ( 1 - y ), x = -sqrt(2) .. sqrt(2),
>                y=-1 .. 1, axes=BOXED, title='Implicit Plot' );
```

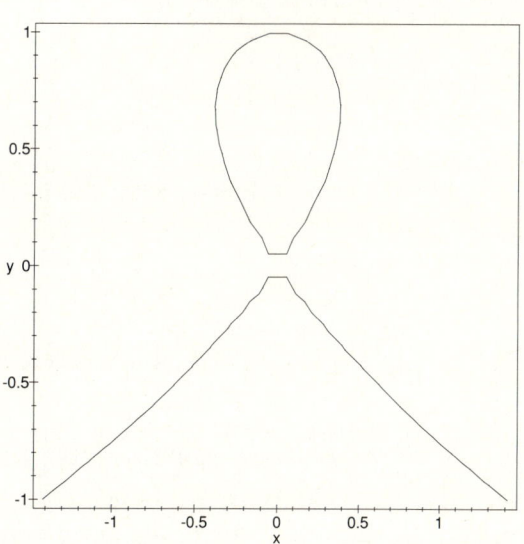

Implicit Plot

. . . .

The fact that the curve obtained from **implicitplot** appears in two pieces is surprising, particularly when it is noticed that the point (0,0), which is immediately seen to satisfy the equation, is not on the curve. The next few examples and Try It! exercises conclude with a good plot of this function.

Try It

It is natural to assume that the problem with the plot in Example 4-7 can be resolved by plotting more points. Investigate this by using the **numpoints=** option to determine the minimum number of points necessary to have the curve plotted as a single piece. Is any other detail lost when this occurs?

Hints:

1. It might be helpful to look at the plot of the individual points. This can be done either by specifying **style=POINT** as an optional argument (see **?plot,options**) or interactively via the icons in the context bar.

2. The default number of points in the plot is 49.

Graphing an implicitly defined function is, in general, a very difficult computational problem. Although Maple does a reasonable job of plotting many implicitly defined functions, the knowledgeable user needs to realize that there may be alternative methods for obtaining the plot of an implicitly defined function.

For a curve like the one in Example 4-7, the classical approach is to express the curve in a parametric form, where $x = ty$ and t is the parameter. Substituting $x=ty$ into the equation for the curve yields $y = 1 - t^2$. Then, $x = ty = t(1 - t^2)$. The constraint $|y| \leq 1$ can be converted into a constraint on the parameter t. This can be done by hand, but it is a nice opportunity to demonstrate the use of Maple with inequalities:

```
> EQS := [ y = 1-t^2, x=t(1-t^2) ];
```

$$EQS := [y = 1 - t^2, x = t(1 - t^2)]$$

```
> solve( subs( EQS, abs(y) )<=1, t );
```

$$\text{RealRange}(-\sqrt{2}, \sqrt{2})$$

Thus, the parametric representation of this curve is $(t(1 - t^2), 1 - t^2)$ for $-\sqrt{2} \leq t \leq \sqrt{2}$. The plot obtained from this representation is

```
> plot( subs( EQS, [ x, y, t = -sqrt(2) .. sqrt(2) ] ),
>        axes=BOXED, title='Parametric Curve' );
```

Parametric Curve

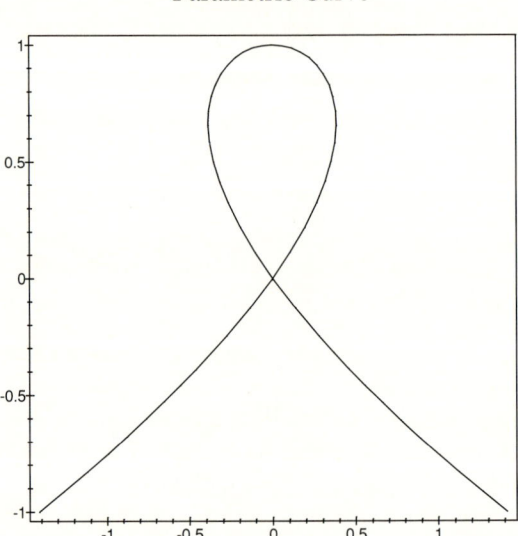

This graph looks much better—it is smooth, contains the origin, and appears in one piece.

Vector-Valued Functions

The graphical representation of a *vector field* can be very useful. For example, the *gradient field* for the function

```
> V := sqrt( x^2+y^2+4 );
```

$$V := \sqrt{x^2 + y^2 + 4}$$

is

```
> gradV := [ diff(V,x), diff(V,y) ];
```

$$gradV := \left[\frac{x}{\sqrt{x^2 + y^2 + 4}}, \frac{y}{\sqrt{x^2 + y^2 + 4}} \right].$$

(See Chapter 6 for a discussion of the **diff** command.) A *normal field* is a vector field that is orthogonal to the gradient field. For example,

```
> normV := [ gradV[2], -gradV[1] ];
```

$$normV := \left[\frac{y}{\sqrt{x^2 + y^2 + 4}}, -\frac{x}{\sqrt{x^2 + y^2 + 4}} \right]$$

The **fieldplot** command is used to plot a vector-valued function. In this case,

```
> G := fieldplot( gradV, x=-2..2, y=-2..2, grid=[8,8],
>                 title='Gradient Field'):
> N := fieldplot( normV, x=-2..2, y=-2..2, grid=[8,8],
>                 title='Normal Field'):
```

These fields can be plotted side by side when the **array** command is used to create an array of plots and then **display** is used to show the composite plot; that is,

```
> display( array( [ G, N ] ) );
```

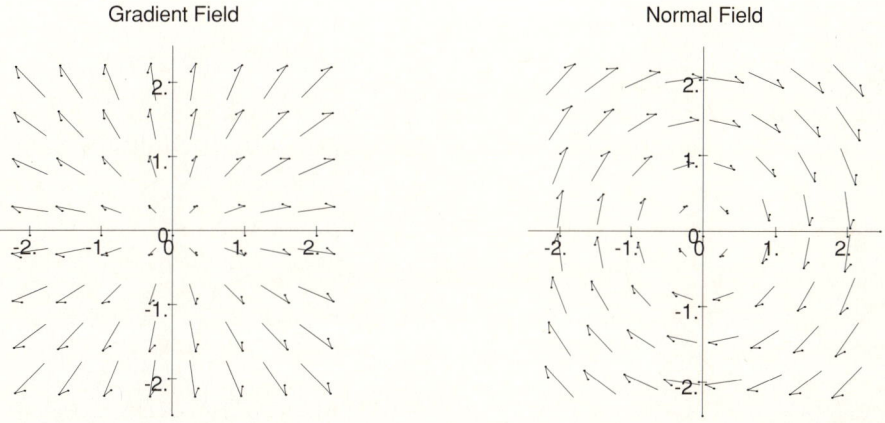

The same idea can be used to create a two-dimensional array, for example, two rows of four columns, of plots.

Try It Create a 3×3 array of plots that confirms the earlier observation that the bandwidth of a filter decreases as the finesse increases. Be sure that each plot is labeled and that the same scaling is used in each plot.

4-3 THREE-DIMENSIONAL PLOTS

Maple can also display a variety of three-dimensional plots, including functions of two variables, implicitly and parametrically defined surfaces, level curves, space curves, and vector fields. In most cases, the Maple command to generate these plots is similar to one of the two-dimensional plot commands. For example, the basic three-dimensional plot command is **plot3d**. The **plots** package contains **implicitplot3d** (corresponding to **implicitplot**) and **animate3d** (for animations) as well as commands for the creation of contour plots (**contourplot** and **contourplot3d**), space curves (**spacecurve**), vector fields (**fieldplot3d**), and plots of objects defined in spherical (**sphereplot**) and cylindrical (**cylinderplot**) coordinates.

EXAMPLE 4-8

Temperature as a Function of Time and Position

The time- and position-varying temperature in a thin rod of unit length in which one end is insulated and the temperature at the other end is held fixed at 0 degrees Celsius is given by $\Theta = 12\,\sin\!\left(\dfrac{\pi x}{2}\right)e^{-t/2}$ for $0 \le x \le 1$, $t > 0$. Plot the temperature (Θ) in the bar for $0 \le t \le 6$. What happens to the heat in this bar as time increases?

SOLUTION

The temperature in the bar is given by

```
> Theta := 12*sin(Pi*x/2)*exp(-t/2);
```

$$\Theta := 12\,\sin\!\left(\frac{1}{2}\,\pi\,x\right)e^{(\,-1/2\ t)}$$

To plot this function, with meaningful labels and title, you could use

```
> plot3d( Theta, x=0..1, t=0..6, axes=BOXED,
>          labels=['position','time','temp'], title='Heat in a Rod' );
```

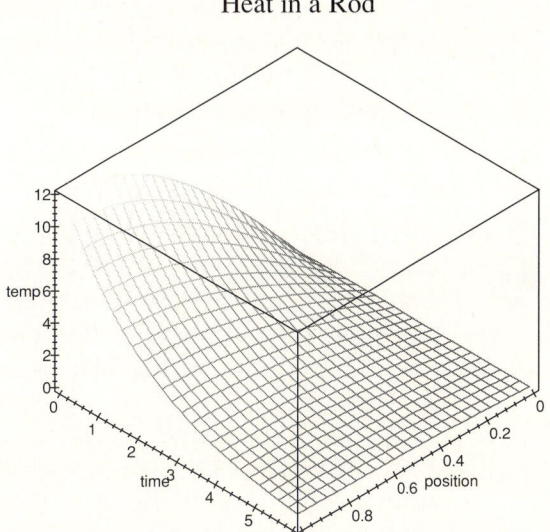

The plot illustrates that the temperature is always zero at $x = 0$. To see that the other end is insulated, note that the surface is flat (in the x direction) so that there is no transfer of heat. Although the temperature initially has some significant variation, by $t = 6$ the temperature is essentially zero. As time continues to increase, the negative exponential in the temperature (and boundedness of the trigonometric term) pushes the temperature toward zero at every point in the bar.

· ·

Note the use of several familiar optional arguments in the **plot3d** command. A full listing of the options can be found in the online help worksheet with keyword **plot3d,option**. Other methods for customizing a three-dimensional plot can be found in the Style, Color, Axes, and Projection menus on the menu bar, on the context bar (see Figure 4-4), and on the pop-up menu that is activated by clicking the right mouse button in the graphics region. For example, the **style=** value of the plot can be changed from the default **HIDDEN** style to a patched surface with contour lines (**PATCHCONTOUR**). Similarly, the **shading=** value of the surface can be selected from **XYZ**, **XY**, **Z**, **ZGREYSCALE**, **ZHUE** (a popular choice for many plots), and **NONE**.

Figure 4-4
Menu bars for three-dimensional graphics regions

Menu bar

Tool bar

Context bar

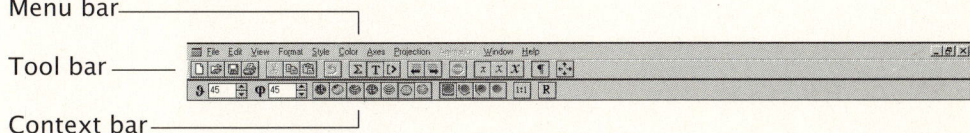

The orientation of the plot can be changed by entering the spherical angles θ and φ in the leftmost box on the context bar or explicitly specifying the angles as an ordered pair in the **orientation=** optional argument, or interactively via the mouse. To use the mouse, position the cursor in the plot region and click the left mouse button. A wireframe of the three-dimensional box will replace the plot. This frame can be rotated by moving the mouse while simultaneously holding down the left mouse button. The plot disappears during the process, and the angles θ and φ are updated in the context bar. Release the mouse button when the desired orientation is achieved. To redraw the plot with the new settings, click the left mouse button on the R icon on the context bar, or select Redraw from the pop-up menu that appears when the right mouse button is pressed.

Try It Interesting views related to this example include the cross-sections with temperature vs. time (**orientation=[0,90]**) and temperature vs. position (**orientation=[90,90]**) and the contour lines in the time vs. position (**orientation=[90,0], style=CONTOUR**) cross-section. Create each of these three plots. Use **display** and **array** to display the three plots side by side.

The examples that conclude this section illustrate a few of the three-dimensional plots that can be created in Maple. The **plot3d**, **plots**, and **updatesR4,graphics** help worksheets contain a more extensive collection of examples.

Implicitly Defined Surfaces

A rendering of the elliptic hyperboloid of one sheet $x^2 + y^2 - z^2 = 4$ is an implicitly-defined surface that is easily obtained as the plot of an implicitly-defined surface:

```
> EllHyp1 := x^2 + y^2 - z^2 = 4;
```

$$EllHyp1 := x^2 + y^2 - z^2 = 4$$

```
> implicitplot3d(EllHyp1, x = -5 .. 5, y = -5 .. 5, z = -5 .. 5,
>                title='3d Implicit Plot', orientation=[60,40] );
```

3d Implicit Plot

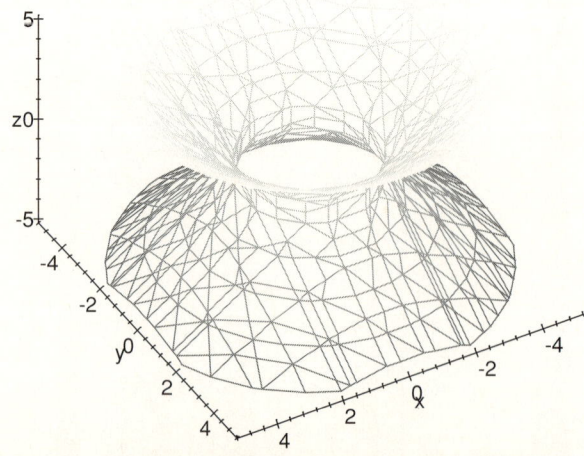

Contour Plots

Topographical, isobar (equal pressure), and isotherm (equal temperature) charts are examples of *contour plots*. These plots, which identify curves along which the given function is constant, are very informative in a variety of engineering disciplines. Examples include the determination of regions that must withstand high pressure concentrations in a submarine or the temperature distribution in a chemical reaction. Maple provides two commands for the creation of contour plots from an expression: **contourplot** and **contourplot3d** (for discrete data, use **listcontplot** and **listcontplot3d**). These commands are similar, but use different algorithms and produce slightly different views of the final contours.

```
> u := 10 * x * exp(-x^2-y^2):
> contourplot( u, x = -2 .. 2, y = -2 .. 2, grid = [49,49],
>                 axes=BOXED, title='Contour Plot');
```

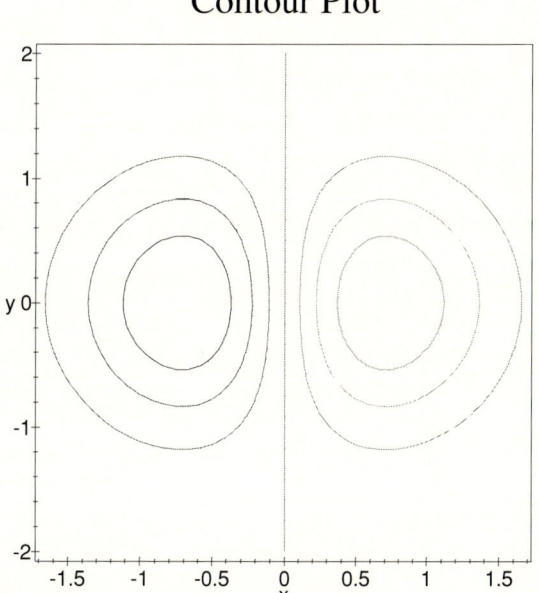

Contour Plot

Try It   Find the on-line help for **contourplot3d**; then use **contourplot3d** to produce a second contour plot of the function *u* defined above. Determine where *u* has its largest and smallest values.

Space Curves

The spacecurve command is used to plot a path through three-dimensional space. For example, given a parametric representation of the spiral path of an object

```
> spiral := [ cos(3*t)/(t+2), sin(3*t)/(t+2), 10-t, t=0..5 ];
```

$$spiral := \left[\frac{\cos(3\,t)}{t+2}, \frac{\sin(3\,t)}{t+2}, 10-t, t=0 .. 5 \right]$$

the path could be viewed using

```
> spacecurve(spiral, axes=BOXED, orientation=[-60,60],
>              title='A Spiral Spacecurve' );
```

A Spiral Spacecurve

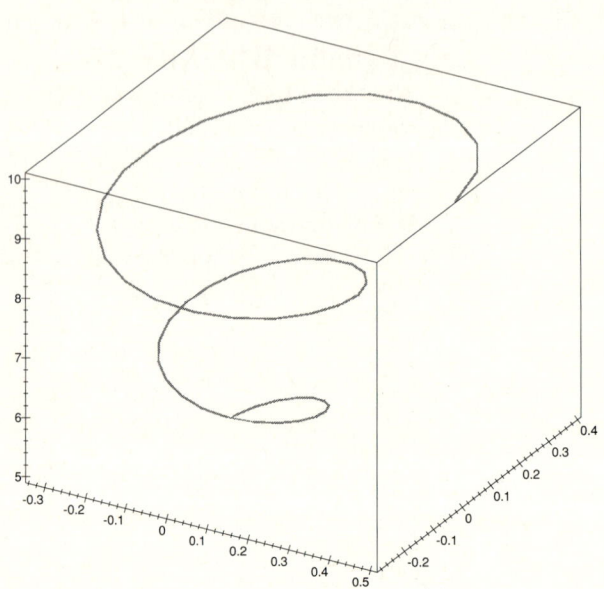

Animation

Three-dimensional animations are useful for the visualization of a function that depends on three variables, for example, $z = f(x, y, t)$.

```
> f := cos(t*x)*sin(t*y);
```

$$f := \cos(tx)\sin(ty)$$

```
> animate3d(f, x = -Pi .. Pi, y = -Pi .. Pi, t = 1 .. 2,
>          title='Waves with Varying Periods');
```

Waves with Varying Periods

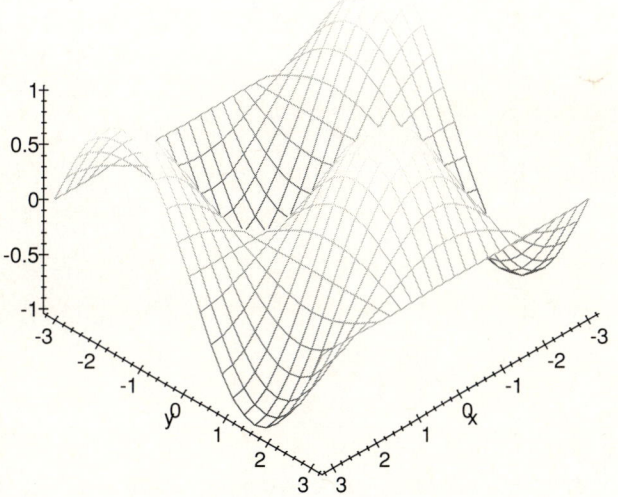

Other Coordinate Systems

The surface defined, implicitly, by $x^2 + y^2 = z^2 y$ could be plotted using **implicitplot3d**, but many of the problems first encountered in Example 4-7 have to be overcome before a good plot is obtained. Cylindrical coordinates present a more natural way to plot this surface. Recall that, in cylindrical coordinates, $x = r\cos(\theta)$ and $y = r\sin(\theta)$ so that $x^2 + y^2 = r^2$. Thus, after cancellation of one power of r, this surface can be written as $r = z^2 \sin(\theta)$. This (that is, $r = F(\theta, z)$) is precisely the form the surface needs to have for use with **cylinderplot**:

```
> cylinderplot( z^2*sin(theta), theta=0..Pi, z=-1..1,
> axes=BOXED, shading=ZHUE, style=PATCHCONTOUR, orientation=[50,73],
> title='Cylindrical Coordinates: r = z^2 * sin(theta)' );
```

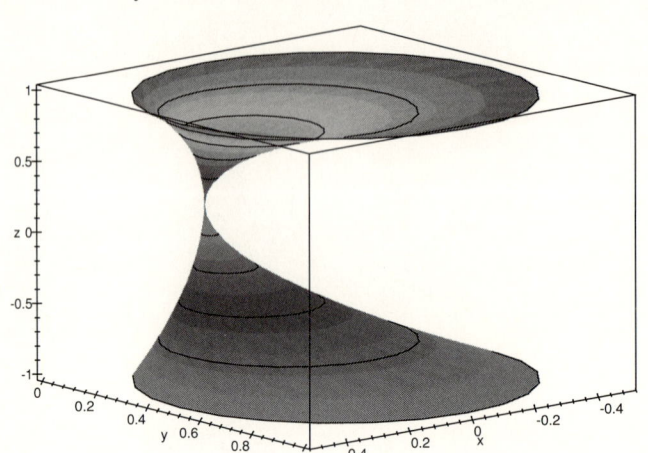

Cylindrical Coordinates: r = z^2 * sin(theta)

Try It

Reconsider the transmission function *T* introduced in the second Try It! exercise in Section 4-2. Originally, animation was used to display this function using two-dimensional plots. Use **plot3d** (and, possibly, **display**) to create a three-dimensional plot of $z = T(x,F)$ and of the plane $z = \dfrac{1}{2}$ in the same plot. How does this picture relate to the individual frames of the animation? Does the animation contain information that is not (directly) available in the three-dimensional plot, or vice versa?

4-4 WORKING WITH DISCRETE DATA

In the first three sections of this chapter, we presented several ways Maple can be used to plot functions and expressions. Engineers also need to be able to visualize and manipulate discrete data points. In fact, you have already encountered this in Application 3 of Chapter 3 (see Table 3-1). In this final section of this chapter, you will learn how to use Maple to plot and manipulate numeric data.

Point Plots

Maple provides a number of ways to plot discrete data. The key is to assemble the data as a list (or set) of ordered pairs [*x*, *y*] or triples [*x*, *y*, *z*]. In two dimensions, the **plot** command is used. To prevent Maple from connecting consecutive points with straight lines, the **style=POINT** optional argument should be specified. In three dimensions, the **pointplot** command from the **plots** package must be used.

EXAMPLE 4-9

Plotting Discrete Points

Plot the five points (0,0), (1,1), (2,1), (2,0), (1,–1).

SOLUTION

Each point is represented as a list with two elements, and the collection of points is saved as an expression sequence:

```
> PTS := [0,0], [1,1], [2,1] ,[2,0], [1,-1];
```

$$PTS := [0, 0], [1, 1], [2, 1], [2, 0], [1, -1]$$

The **plot** command can accept either a list or set of points as input.

```
> plot( [ PTS ], style=POINT, symbol=BOX, title='Five Points' );
```

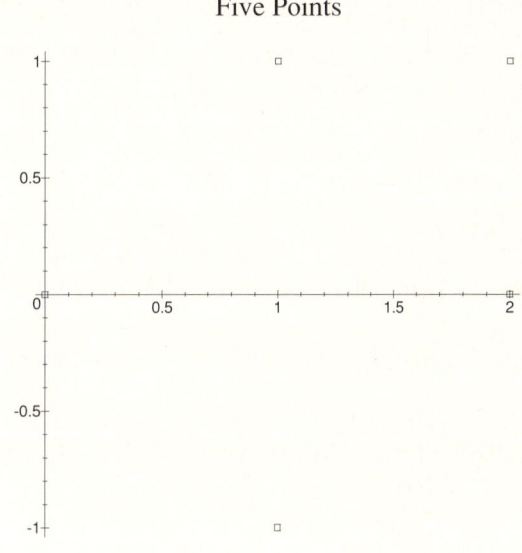

Five Points

· ·

Try It

The five points in Example 4-9 are the vertices of a pentagon. Modify the solution to Example 4-9 to create a plot of this pentagon.

Least Squares Fit to Data

In addition to simply plotting a set of data points, it is sometimes desirable to find the best straight line (or quadratic, exponential, trigonometric, or other function) that can be associated with the given data. The mathematical name associated with this process is *least-squares fit*. The **fit** command from the **stats** package is used for computing least-squares fits to a set of data points.

EXAMPLE 4-10 **The Best Straight Line Fit to Data**

Find, and plot, the best straight line fit ($y = mx + b$) to the four data points (10,3), (15,4), (17,5), (19,6).

SOLUTION

The biggest obstacle in this problem is deciding how to represent the data points. The data for **fit** must be two separate lists of numbers, whereas **plot** accepts a single list of points. Since only four data points are in this problem, it would not be too difficult to construct the different forms of the data manually, but this would not be practical if there were dozens, hundreds, or thousands of data points. The simplest, and most general, solution is to enter the data as two separate lists for **fit** and use the **zip** command to create the composite list for **plot**.

The basic syntax for **zip** requires three arguments. The first argument is a binary function, and the other two arguments are lists; the output is the list obtained by applying the function to consecutive pairs of values in the two lists. (For additional details and examples, consult the help worksheet for **zip**.) For example, if the x- and y-coordinates of the data points are entered as lists

```
> X := [ 10, 15, 17, 19 ];
```

$$X := [10, 15, 17, 19]$$

```
> Y := [ 3, 4, 5, 6 ];
```

$$Y := [3, 4, 5, 6]$$

then the list of points needed for **plot** can be created using **zip** with a first argument that pairs together corresponding elements of the lists **X** and **Y**:

```
> PTS := zip( (x,y)->[x,y], X, Y );
```

$$PTS := [[10, 3], [15, 4], [17, 5], [19, 6]]$$

The plot of these points is now simple:

```
> P := plot( PTS, x=0..20, y=0..8, style=POINT, color=BLUE ):
> display( P, title='Data Points' );
```

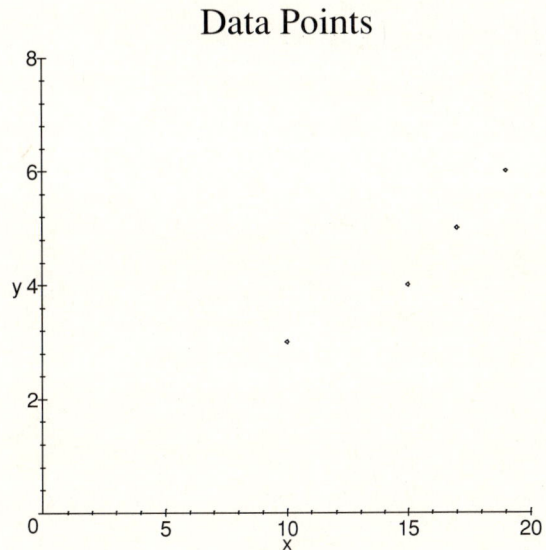

Data Points

Note that this plot is assigned to a name (**P**) so that it can be used later to create the composite plot that includes the best linear fit to this data.

The last step in preparing to use **fit** is to load the **stats** package.

```
> with( stats );
```

[*anova, describe, fit, importdata, random, statevalf, statplots, transform*]

The syntax for **fit** is easy to understand, but it is difficult to describe. The arguments come in two distinct pieces: the description of the function to which the data is to be fitted and the data points. The description of the function requires the general formula for the function (here, $y = mx + b$) and sets of the variables ($\{x, y\}$) and parameters ($\{m, b\}$) in this function. It is suggested that the general form be copied from the on-line help (**?stats,fit**) and modified for the current needs.

```
> SL := fit[leastsquare[ {x,y}, y=m*x+b, {m,b} ] ] ( [ X, Y ] );
```

$$SL := y = \frac{58}{179} x - \frac{79}{179}$$

This line is easy to plot.

```
> P1 := plot( rhs(SL), x=0 .. 20 ):
```

To see how the best-fit line compares with the data, the two plots can be superimposed.

```
> display( { P, P1 }, title='Data Points and Best Linear Fit' );
```

Data Points and Best Linear Fit

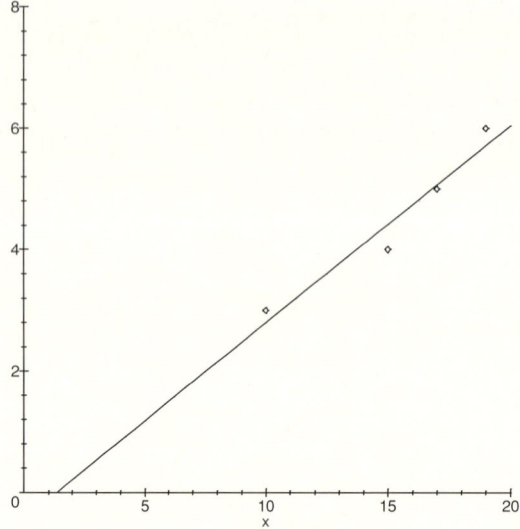

Note that the line does not pass through any of the data points. This is a common characteristic of least-squares fits to data.

. . . .

Try It
⚠

Repeat the previous steps to find the best quadratic fit to the same set of data. Plot the data points, the best linear fit, and the best quadratic fit all on the same set of axes.

To conclude this introduction to plotting and data fitting, you will explore the relationship between reflectivity, *R*, and finesse, *F*, using the data-fitting techniques just introduced.

EXAMPLE 4-11   ## Visualizing the (FWHM) Bandwidth

The bandwidth of a filter was defined in Application 4 as the FWHM of a plot of the transmitted light vs. frequency (see Figure 4-3). The relationship between finesse and those frequencies corresponding to the "half-maximum" condition is given by $T(x, F) = \dfrac{1}{2}$. Use the graph of this implicitly defined function to estimate the bandwidth (FWHM) of the filter when the finesse is between 5 and 50. Conclude the data collection stage by plotting the bandwidth vs. finesse.

SOLUTION

Recall that the transmission function is

```
> F := 'F':
> T := 1/(1+(2*F*sin(Pi*x)/Pi)^2);
```

$$T := \frac{1}{1 + 4\dfrac{F^2 \sin(\pi x)^2}{\pi^2}}$$

The relationship between finesse and the half-maximum frequencies is contained in the implicitly defined function $T(x, F) = \dfrac{1}{2}$. The graph of this function, for finesse between 5 and 50, is

```
> implicitplot( T=1/2, x=0.9..1.1, F=5..50,
>                  title='Finesse vs. Frequency' );
```

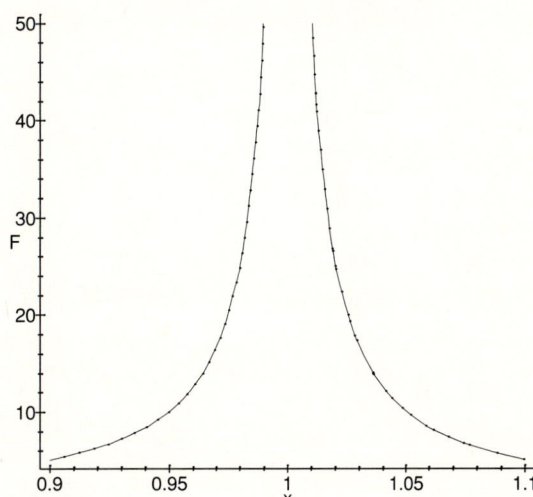

This plot can be used to gather data points that should shed more light on the relationship between the bandwidth and finesse *F*. It will be simplest to keep two separate lists: one (**Flist**) for the finesse values and one (**Blist**) for the bandwidths. The bandwidths are estimated from the plot by computing the FWHM for several different finesse levels. The data collection process is a little tedious, but not too bad as your mouse skills improve.

```
> Flist := [ 5, 10.1, 15.01, 20.19, 30.01, 40.11, 49.93 ];
```

$$Flist := [5, 10.1, 15.01, 20.19, 30.01, 40.11, 49.93]$$

```
> Blist := [ 1.096-0.9016, 1.049-0.9502, 1.033-0.966,
  1.023-0.9745, 1.016-0.9818, 1.012-0.9867, 1.01-0.9891 ];
```

$$Blist := [.1944, .0988, .067, .0485, .0342, .0253, .0209]$$

The first step is to plot the data and look for general patterns that can be investigated in more detail. The data are converted to the form suitable for plotting using **zip**, and the plot is assigned to a name (**P**) for later reference.

```
> PTS := zip( (x,y) -> [x,y], Flist, Blist );
```

$$PTS := [[5, .1944], [10.1, .0988], [15.01, .067], [20.19, .0485],$$
$$[30.01, .0342], [40.11, .0253], [49.93, .0209]]$$

```
> P := plot( PTS, style=POINT, labels=['Finesse', 'FWHM'], color=RED ):
> display( P, title='Bandwidth vs. Finesse (data points)' );
```

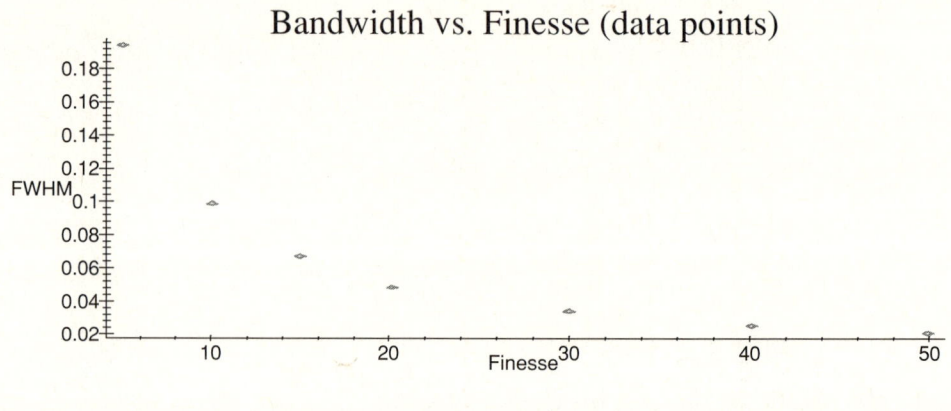

The fact that the bandwidth decreases toward zero with increasing finesse suggests a reciprocal relationship between finesse and bandwidth, but is bandwidth proportional to $\dfrac{1}{F}$, $\dfrac{1}{\sqrt{F}}$, or something more complicated than a reciprocal power?

<table>
<tr><td>**EXAMPLE 4-12**</td><td></td></tr>
</table>

An Inverse Relationship Between Bandwidth and Finesse

A visual check for the relationship between bandwidth and finesse is based on the observation that if the bandwidth is proportional to a reciprocal power of finesse, that is, $FWHM = \alpha F^{-p}$, then the finesse is proportional to bandwidth raised to the power $-\dfrac{1}{p}$. Construct and plot the $FWHM^{-1/p}$ vs. F for $p = \dfrac{1}{2}$, $p = 1$, and $p = 2$. Which of these powers is the best candidate for the relationship between bandwidth and finesse?

SOLUTION

The **zip** command can perform the data transformations that are needed

```
> PTS1 := zip( (x,y) -> [x,y^(-2)], Flist, Blist ):
> PTS2 := zip( (x,y) -> [x,y^(-1)], Flist, Blist ):
> PTS3 := zip( (x,y) -> [x,y^(-1/2)], Flist, Blist ):
```

The output of the previous commands is suppressed because the graphs are expected to provide all necessary information for this comparison. The graphs are easily created using **plot**. Note that the linearity of the curve will be easier to determine if the points are connected.

```
> PP1 := plot( PTS1, labels=['F','FWHM'], color=GREEN, title='p = 1/2' ):
> PP2 := plot( PTS2, labels=['F','FWHM'], color=RED, title='p = 1' ):
> PP3 := plot( PTS3, labels=['F','FWHM'], color=BLUE, title='p = 2' ):
```

The plots can be superimposed in one graph, but the scale for $p = \dfrac{1}{2}$ is sufficiently large that the other two plots are essentially invisible. For this reason it is preferable to display the three plots side by side.

```
> display( array( [ PP1, PP2, PP3 ] ) );
```

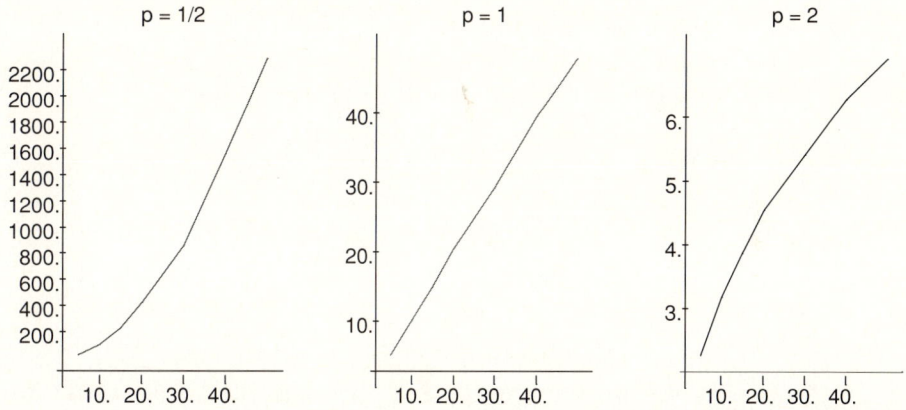

These plots suggest that, of these three choices of p, the bandwidth is most likely to be inversely proportional to finesse, as was expected from the definition of finesse.

· ·

EXAMPLE 4-13   **The Best Reciprocal Fit to the FWHM Data**

Find the best reciprocal fit to the bandwidth and finesse data. Display the data points and the best reciprocal fit in the same graph.

SOLUTION

The general form for a reciprocal fit is $B = \dfrac{a}{F}$. The least-squares fit of this form is found to be:

```
> recipfit := stats[fit,leastsquare[ {F,B}, B=a/F], {a} ]
                 ( [Flist,Blist] );
```

$$recipfit := B = \frac{1.024940114}{F}$$

Even though this is the best fit, this is not a guarantee that the curve is close to the data points. A combined plot of the data points and best fit function is usually the best way to look for agreement between the data points and best fit function.

```
> P1 := plot( rhs(recipfit), F=5..50 ):
> display( { P, P1 }, title='Quality of Best Reciprocal Fit' );
```

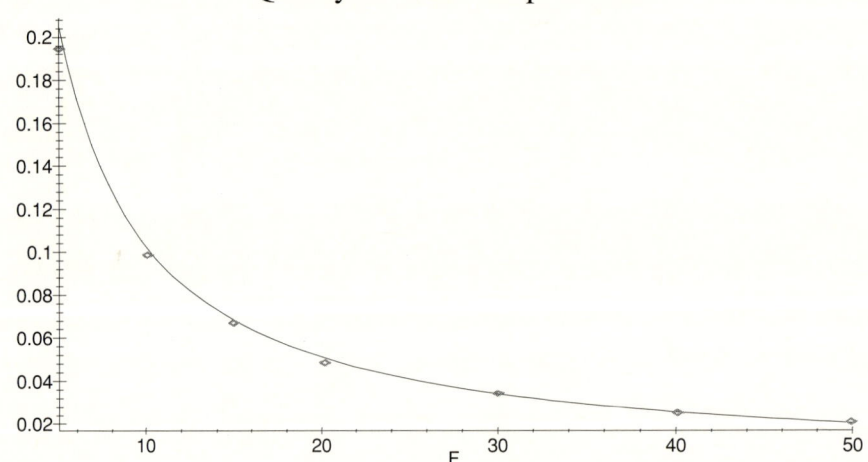

Quality of Best Reciprocal Fit

That looks like a very good fit. And there's more. The coefficient in the fit is amazingly close to 1. Are the bandwidth and finesse exactly reciprocals of one another?

. . . .

Try It Repeat the data-fitting computations looking for fits of the form $B = \dfrac{a}{F} + b$

and $B = \dfrac{a}{F^2}$ and $B = \dfrac{a}{F} + b\,F + c$. How do these compare with the best reciprocal fit?

SUMMARY

This chapter showed you how to plot many types of engineering functions and data. Examples of two- and three-dimensional plots, as well as parametric, implicit, field (vector), spacecurve, and contour plots, and animations were provided. The appropriate use of titles, labels, colors, styles was discussed as an essential part of any effective presentation of engineering data. In addition, the chapter introduced the concept of data fitting and least squares approximations to data. The optical filter application incorporated many of these ideas while investigating the relationships between the bandwidth, reflectivity, and finesse of an optical interference filter.

Complete information about Maple's plotting commands and their options can be found in the on-line help. Good places to start a search for information about Maple's plotting commands are the help worksheets for **plot**, **plot3d**, and **plots**. Specialized routines for the least squares and discrete data analysis can be found in the **stats** package.

Key Words

animation	normal field
aspect ratio	parametric curve
contour plot	space curve
gradient field	vector field
implicitly defined function	vertical asymptote
least-squares fit	

Maple Commands

abs	**plot**, options:
array	**axes**, **color**, **discont**,
evalf	**labels**, **linestyle**,
plot	**numpoints**, **orientation**,
plot3d	**scaling**, **shading**, **style**,
rhs	**symbol**, **title**, **view**
solve	**plots** package:
with	**display**
zip	**animate(3d)**, **contourplot(3d)**,
	cylinderplot, **fieldplot(3d)**,
	implicitplot(3d), **spacecurve**,
	sphereplot
	stats package:
	fit, **leastsquare**

References

1. Yariv, A., *Optical Electronics in Modern Communications*, New York: Oxford University Press, 1997.
2. Saleh, B.E.A. and Teich, M.C., *Fundamentals of Photonics*, New York: John Wiley & Sons, Inc., 1991.

Problems

1. Plot the following functions on the specified domains. Select optional arguments so that the final plot clearly illustrates the interesting features of the function. Be certain to include labels and a title.

    (a) Plot $f(x) = e^{-(x-2)^2} \sin(\pi x)$ and $g(x) = e^{-(x-2)^2} \sin^2(\pi x)$ for the first two periods of the trigonometric terms.

    (b) Plot $u(x, y) = x\sin(y) - y\cos(x)$ for $0 \leq x, y \leq 4\pi$.

    (c) Plot $F(u, v) = \dfrac{x^2 - y^2}{x^2 + y^2}$ for $-1 \leq x, y \leq 1$.

    Note: Look at the contour lines. What is the interesting point for this function?

    (d) Plot $v_1(\theta) = \dfrac{\sin(\theta)}{\theta}$ and $v_2 = \dfrac{1 - \cos(\theta)}{\theta}$ for $|\theta| \leq 5\pi$

2. Describe the closed curve defined implicitly by

    $$4x^2 + 2\sqrt{3}\,xy + 2y^2 + 10\sqrt{3}\,x + 10y = 5.$$

    Is the origin $(0,0)$ inside the region bounded by this curve?

3. Determine, both graphically and analytically, the percent error in the reflectivity that is needed to ensure that the finesse is controlled to within (plus or minus) 3% of 50.

4. The ratio of light that passes through a filter to that which enters the filter is $T = \dfrac{1}{1 + \dfrac{4F^2 \sin(\pi x)^2}{\pi^2}}$ where x is proportional to the frequency of the light and F is the finesse.

    (a) Plot the transmitted light for a filter with a finesse $F = 20$ and for $0.9 \leq x \leq 1.1$.

    Note that the transmission is greatest when $x = 1$ in this plot. More generally, $T = 1$ whenever x is an integer. The bandwidth of a filter is typically determined by its full width at half maximum (FWHM). That is, the maximum transmission coefficient is 1 (100%) when $x = 1$, and half of the signal is transmitted ($T = \dfrac{1}{2}$) when $x = 0.974$ and again when $x = 1.03$; the difference between these "frequencies" is the FWHM for $F = 20$.

    (b) Use the **animate** command to determine whether larger or smaller values of F are needed to produce a filter with a narrower bandwidth. To simplify the identification of the frequencies used to compute the FWHM, also plot the horizontal lines $T = \dfrac{1}{2}$ in each frame of the animation.

5. (a) Beginning with the optical filter transmission function T(x,F) given in problem 4, use Maple to symbolically solve for the bandwidth (B) in terms of finesse (F). Does B change from peak to peak in this example?

    (b) Repeat part (a) for the high-finesse case by replacing $\sin(\pi x)$ by πx in T(x,F). What advantages are obtained from making this approximation? How small does x need to be to ensure that the errors arising from this approximation are not too large? (*Hint:* Use a plot.)

    (c) Plot B vs. F on the same graph using your results from (a) and (b). Over what range of F does the approximation incorporated in (b) have an error of less than 1%?

    (d) Plot T vs. x for $0 < x < 3$ for $0.5 < F < 50$ using the **animate** function. Do you see a problem with the bandwidth definition for low F? (*Hint:* It might be more useful to look at the animation with decreasing values of F.)

6. In Example 4-11, we (correctly) guessed that there is a reciprocal relationship between the finesse and bandwidth of a filter. A log–log plot can be used to obtain similar information about a set of data. The basic idea is that if $F = B^\alpha$, for some constant α, then $\log(F) = \alpha\log(B)$. That is, the graph of $\log(B)$ vs. $\log(F)$ will be a straight line with slope α.

    Confirm the results of Example 4-13 by creating a log–log plot of the data points used in that example.

7. Perform a least-squares fit of the wind tunnel and CFD data in Application 3 (Chapter 3). Graph the data points, the lift-to-drag function found in Step 4, and the best-fit solution. How do these fits compare with the estimate found in Step 4 of the five-step solution? What are the drag coefficients and thrust requirements when $C_L = 0.5060$?

8. Use **contourplot** (or **contourplot3d**) to plot level curves of the following two equations:

$$\left(u - \frac{r}{1+r}\right)^2 + v^2 = \frac{1}{(1+r)^2}$$

$$(u-1)^2 + \left(v - \frac{1}{x}\right)^2 = \frac{1}{x^2}$$

Plot lines having constant r = 0, 0.5, 1, 2, 5, 10 in the u–v plane. Do the same for lines having constant x = 0, 0.5, –0.5, 1, –1, 2, –2, 5, –5, 10, and –10. This type of plot, called a Smith chart, is used by engineers to describe more complicated relationships between various quantities in microwave engineering. For example, they may describe a transformation between reflection coefficient and normalized impedance in a coaxial cable being used as a transmission line by electrical engineers analyzing a communication channel.

9. The **textplot** (and **textplot3d**) commands (from the **plots** package) can be used to insert labels in a two- or three-dimensional plot. Find the online help for these commands, then use them to identify the two curves plotted in Example 4-2.

10. Consider any periodic function of time f with period T, returning once each cycle to any selected time reference. In other words, $f(t + T) = f(t)$, where T is called the period of the periodic function. This period is the minimum time it takes the function to duplicate itself. Such a function may be represented by a group of purely sinusoidal functions, consisting of a fundamental frequency and its harmonics. Fourier analysis allows us to find the "weighted" coefficients of each of the sinusoidal terms in such a way that, after adding them together, they approximate the original periodic function. For example,

    let $\theta = \dfrac{2 \pi t}{T}$, where t is the time and T is the period; θ is in radians.

    Define:

    $$f_{2 n-1}(\theta) = \frac{1}{2} + \frac{2}{\pi} \sum_{k=1}^{n} (-1)^{k-1} \frac{\cos((2 k - 1) \theta)}{2 k - 1} \quad \text{for } n = 1, 2, \ldots$$

    Graph, in one plot, the functions f_1, f_3, and f_5 for at least two periods. Then plot f_{25} on a separate plot. What periodic function is being repereseted by this group of sinusoidal functions?

11. Plot the functions $y = \tan(x)$ and $y = x$ on a single graph. Be sure to choose the domain so that the graph contains the first five positive values of x for which $\tan(x) = x$. Use the interface to identify the first five positive values of x for which $\tan(x) = x$. Compare these results with the values found in Example 3-17.

12. Create, in one plot, a graph of the CFD data from Application 3 (Chapter 3) and the quadratic function relating the coefficients of lift and drag that is given in Step 4.

13. (a) The curve $x^2 = y^2(1 - y)$ was plotted as an implicitly defined function and parametrically in Section 4-2. The points with $t = \sqrt{2}$ and $t = -\sqrt{2}$ correspond to the two endpoints of the curve. Find the value(s) of the parameter t when the curve passes through the origin.

    (b) Use **animate** to show how the curve is traced out as t increases from $-\sqrt{2}$ to $\sqrt{2}$.

14. Verify that the gradient and normal fields for $V = \sqrt{x^2 + y^2 + 4}$ are orthogonal by superimposing the plots of the two vector fields on top of one another. (Use different colors to distinguish the two vector fields.)

5 Manipulating Expressions with Maple

Water Quality This chapter provides a practical example showing how environmental engineers use criteria such as the dissolved oxygen level, temperature, and concentration of various organic and inorganic compounds to assess water quality. You will use the Streeter–Phelps equation to model the dissolved oxygen (DO) sag curve for a stream subjected to pollutants. The DO sag curve essentially displays the level of water quality as a function of distance from the pollution site. Two important pieces of information provided by this curve are the minimum level of dissolved oxygen in the stream and how far downstream from the site it occurs. Although this problem could be solved using your current Maple skills, the new commands and techniques introduced in this chapter will provide you with greater control over the specific appearance and form of your results.

INTRODUCTION

You now should be feeling more comfortable with Maple's user interface, basic commands for symbolic, numeric, and graphic manipulations, and online help. You should also be seeing ways Maple can be used to solve engineering problems. Although Maple usually provides results in the form you expect, there are times when the output needs some additional massaging before it has the desired form. For instance, in the Try It! exercise that immediately precedes Example 3-5, the fact that the radius of a cone is positive is used to simplify the formula for the surface area of a cone, and in Example 3-8, the list of perfect squares and cubes is more easily interpreted when the integers are sorted. In this chapter, you will learn how to use Maple to manipulate expressions, functions, sets, lists, and other objects into the precise form and type that you require. Application 5 illustrates how an environmental engineer might analyze the quality of water to ascertain whether fish life can be sustained.

The **simplify** and **collect** commands, introduced in Chapter 3, are examples of Maple commands that modify the appearance of a Maple object. This chapter continues the discussion of **simplify**. This chapter also discusses some other commands for performing precise manipulations on a Maple object: **normal**, **factor**, **expand**, and **combine**; the command names indicate the specific operation performed on the argument. The **assume** facility, also introduced in Chapter 3, provides additional information about a specific Maple object. The discussion of the **assume** facility and related techniques begins in Section 5-1 and continues throughout the chapter.

Note that although most of the initial examples involve expressions, each of the commands discussed in this section can be applied to an equation, range, or function, or to a list or set of these objects. Moreover, in an attempt to focus on the action of each command, only a few of the examples and Try It! exercises in this chapter have a direct engineering application. As usual, the online help should be consulted for a complete description of the required and optional arguments.

5-1   USING `simplify`, SIDE RELATIONS, AND `assume`

The most commonly used simplification command is **simplify**, which applies a set of *simplification rules* to Maple expressions. A typical simplification is the use of the identity $\sin^2(x) = 1 - \cos^2(x)$ in trigonometric terms.

EXAMPLE 5-1

Simplifying Trigonometric Expressions

Simplify, first by hand and then using Maple, each of the following three expressions: $y_1 = \cos(2x) - \sin^2(x)$, $y_2 = \cos(2x) - \cos(x)$, and $y_3 = 1 - \cos^2(2x)$.

SOLUTION

The first expression can be simplified using the double-angle formula for cosine to find that

$$y_1 = \cos(2x) - \sin^2(x) = \cos^2(x) - \sin^2(x) - \sin^2(x) = \cos^2(x) - 2\sin^2(x)$$

If $\cos^2(x)$ is replaced with $1 - \sin^2(x)$, then $y_1 = 1 - 3\sin^2(x)$, and if $\sin^2(x)$ is replaced with $1 - \cos^2(x)$, then $y_1 = 3\cos^2(x) - 2$. Maple's **simplify** command returns

```
> restart;
> EXPR1 := cos(2*x) - sin(x)^2:
> EXPR1 = simplify( EXPR1 );
```

$$\cos(2x) - \sin(x)^2 = 3\cos(x)^2 - 2$$

This result indicates that Maple uses $\sin^2(x) = 1 - \cos^2(x)$, not $\sin^2(x) + \cos^2(x) = 1$, as the basis for this simplification rule.

Manual simplification of the second expression gives

$$y_2 = \cos(2x) - \cos^2(x) = \cos^2(x) - \sin^2(x) - \cos^2(x) = -\sin^2(x)$$

Maple goes one step further and applies $\sin^2(x) = 1 - \cos^2(x)$ to obtain its result:

```
> EXPR2 := cos(2*x) - cos(x)^2:
> EXPR2 = simplify( EXPR2 );
```

$$\cos(2x) - \cos(x)^2 = -1 + \cos(x)^2$$

Manual simplification of the third expression yields $y_3 = 1 - \cos^2(2x) = \sin^2(2x)$. The double-angle formula for sine could be applied, but this would introduce more terms in the expression. Maple's simplified form is

```
> EXPR3 := 1 - cos(2*x)^2:
> EXPR3 = simplify( EXPR3 );
```

$$1 - \cos(2x)^2 = 1 - \cos(2x)^2$$

. . . .

One lesson to be learned from Example 5-1 is that it can be quite difficult to define "simplified form," even for the relatively elementary examples so far considered. When using Maple's simplification commands, you need to remember that Maple "knows" only what it has been told about the problem. It may be necessary for you, the user, to manually rewrite an answer in the form most suitable for your purposes.

An alternative is to specify, as a second argument to **simplify**, a set of *side relations* containing specific simplifications to be applied during simplification. (For a full description of the uses and specification of side relations, see the **simplify[siderel]** help worksheet.)

EXAMPLE 5-2

Customized Simplifications

Repeat the simplifications in Example 5-1 when $\cos^2(x)$ is replaced with $1 - \sin^2(x)$.

SOLUTION

The side relation necessary for these problems is

> `SIDEREL := { cos(x)^2 = 1 - sin(x)^2 }:`

Now the three expressions simplify to

> `EXPR1 = simplify( EXPR1, SIDEREL );`

$$\cos(2x) - \sin(x)^2 = 1 - 3\sin(x)^2$$

> `EXPR2 = simplify( EXPR2, SIDEREL );`

$$\cos(2x) - \cos(x)^2 = -\sin(x)^2$$

> `EXPR3 = simplify( EXPR3, SIDEREL );`

$$1 - \cos(2x)^2 = 4\sin(x)^2 - 4\sin(x)^4$$

The first two results are obviously consistent with the results of Example 5-1. The last result requires some additional thought. Note that the right-hand side can be factored:

$$4\sin^2(x)(1 - \sin^2(x)) = 4\sin^2(x)\cos^2(x) = (2\sin(x)\cos(x))^2 = \sin(2x)^2$$

This suggests that Maple applied the double-angle formula for sines in addition to the simplification given as a side relation.

. . . .

Try It

Find the simplest expression that is equivalent to

$$\sin^7(x) + \sin^5(x)\cos(x) + \sin^5(x)\cos^2(x) - \sin^3(x)\cos^3(x).$$

_____

A second transformation applied by **simplify** is the replacement of a radical of the form $(x^n\, y)^{\left(\frac{m}{n}\right)}$ with $x^m\, y^{\left(\frac{m}{n}\right)}$, when $x > 0$, and m and n are positive integers. If the extra conditions for x, m, and n are not all satisfied, Maple will not apply this particular simplification rule. By default, Maple assumes that all names are complex-valued unless specified otherwise. The **assume** command, introduced in Chapter 3, is one method that can be used to provide specific information about a name. Two alternative methods for achieving similar results are the **assume=** and **symbolic** optional arguments to **simplify**.

EXAMPLE 5-3   **assume and simplify**

Simplify $\left(x\, y^4 \right)^{\left(\frac{3}{q+1} \right)}$ using the optional arguments to **simplify** and using Maple's **assume** facility. Which method provides more information about the conditions under which this expression can be simplified?

SOLUTION

First, note that the simplified form is expected to be $x^{\left(\frac{3}{q+1} \right)} y^{\left(\frac{12}{q+1} \right)}$, but that **simplify** with no additional information only partially reduces the expression to this form:

```
> restart;
> EXPR := (x*y^4)^(3/(q+1)):
> EXPR = simplify( EXPR );
```

$$\left(x\, y^4 \right)^{\left(\frac{3}{q+1} \right)} = \left(x^3\, y^{12} \right)^{\left(\frac{1}{q+1} \right)}$$

When the **symbolic** option is specified, Maple essentially turns off all checks for validity:

```
> simplify( EXPR, symbolic );
```

$$x^{\left(\frac{3}{q+1} \right)} y^{\left(\frac{12}{q+1} \right)}$$

This is the expected simplified form of the expression, but this approach does not give any indication of the conditions under which this transformation is valid. The **symbolic** option should be used very sparingly, and all results should be checked to be sure they are valid for your intended use.

The **assume=** optional argument specifies a single property, such as **positive** or **real**, for all names in the expression. When x, y, and q are all positive, the expression reduces to

```
> simplify( EXPR, assume=positive );
```

$$x^{\left(\frac{3}{q+1} \right)} y^{\left(\frac{12}{q+1} \right)}$$

The **assume=** option is preferable to the **symbolic** option in that you now know that the simplification is valid when x, y, and q are all positive.

In many problems, including this one, it is not desirable to have all names of the same type. The **assume** command gives you individual control over each name (see also Chapter 3). The conditions given for this example are certainly satisfied when $q > -1$, $y > 0$, and $x > 0$:

```
> assume( q>-1, y>0, x>0 );
> about( [q,x,y] );
```

```
[q, x, y]:
    is used in the following assumed objects
    [x] assumed RealRange(Open(0),infinity)
    [y] assumed RealRange(Open(0),infinity)
    [q] assumed RealRange(Open(-1),infinity)
```

The simplification that results under these assumptions is

```
> simplify( EXPR );
```

$$x\mathord{\sim}^{\left(\frac{3}{q\mathord{\sim}+1}\right)}\,y\mathord{\sim}^{\left(\frac{12}{q\mathord{\sim}+1}\right)}$$

Observe that the ability to independently control the property of each name with **assume** provides much greater flexibility than do either of the optional arguments to simplify. For this reason, you are strongly encouraged to use the commands in the **assume** facility (**assume**, **about**, and **additionally**) whenever it is not appropriate for Maple to assume that a name is a general complex number.

. . . .

Try It

Repeat the simplification of the expression in Example 5-3 for each combination of two assumptions for *x*, *y*, and *q*. Explain your results.

Additional information about the two simplifications discussed in this section can be found in the online help under the topics **simplify[trig]** and **simplify[radical]**. The complete list of simplifications known to Maple is quite extensive; some of the more common transformations are described under the help topics **simplify[ln]**, **simplify[power]**, **simplify[sqrt]**, **simplify[piecewise]**, and **simplify[RootOf]**.

5-2 USING normal

The **normal** command returns a *rational expression* in which the numerator and denominator are *relatively prime* (that is, they contain no common factors). Even though the algorithms used by **normal** and **simplify** are completely different, in many cases they yield the same results.

Normalized expressions are typically returned with the numerator and denominator in their factored form; this default format can be overridden by including the string **expanded** as an (optional) second argument to **normal**.

EXAMPLE 5-4   ## Comparison of `simplify` and `normal`

Compare the results of simplifying and normalizing the expressions

$$\frac{x^{10}-1}{x^2-1}, \quad \frac{e^{(10x)}-1}{e^{(2x)}-1} \quad \text{and} \quad \frac{\sin^{10}(\theta)-1}{\sin^2(\theta)-1}.$$

SOLUTION

```
> restart;
> EXPR1 := (x^10-1)/(x^2-1);
```

$$EXPR1 := \frac{x^{10}-1}{x^2-1}$$

The second and third expressions are easily generated from the first. The only subtle point is to force the simplification of exponentials without fully simplifying the exponential expression.

```
> EXPR2 := subs( x=exp(x), EXPR1 );
```

$$EXPR2 := \frac{(e^x)^{10}-1}{(e^x)^2-1}$$

```
> EXPR2 := simplify( EXPR2, exp );
```

$$EXPR2 := \frac{e^{(10x)}-1}{e^{(2x)}-1}$$

```
> EXPR3 := subs( x=sin(theta), EXPR1 );
```

$$EXPR3 := \frac{\sin(\theta)^{10}-1}{\sin(\theta)^2-1}$$

The simplified and normalized expressions for each of the three expressions are easily obtained. A summary of the results, with comments, follows:

```
> EXPR1; simplify( EXPR1 ); normal( EXPR1 );
```

$$\frac{x^{10}-1}{x^2-1}$$
$$x^8 + x^6 + x^4 + x^2 + 1$$
$$x^8 + x^6 + x^4 + x^2 + 1$$

Both **simplify** and **normal** are able to factor the numerator and denominator and cancel common factors. Note that, technically, the removal of the singularities at $x = 1$ and $x = -1$ does change some properties of the expression.

```
> EXPR2; simplify( EXPR2 ); normal( EXPR2 );
```

$$\frac{e^{(10\,x)} - 1}{e^{(2\,x)} - 1}$$

$$e^{(8\,x)} + e^{(6\,x)} + e^{(4\,x)} + e^{(2\,x)} + 1$$

$$\frac{e^{(10\,x)} - 1}{e^{(2\,x)} - 1}$$

The results for this expression look different because **normal** is designed for use with rational functions with polynomial numerators and denominators.

```
> EXPR3; simplify( EXPR3 ); normal( EXPR3 );
```

$$\frac{\sin(\theta)^{10} - 1}{\sin(\theta)^2 - 1}$$

$$5 - 10\cos(\theta)^2 + 10\cos(\theta)^4 - 5\cos(\theta)^6 + \cos(\theta)^8$$

$$\sin(\theta)^8 + \sin(\theta)^6 + \sin(\theta)^4 + \sin(\theta)^2 + 1$$

These results look quite different. The absence of sine from the simplified result is reminiscent of the comments in Example 5-1 concerning the "simplification" of powers of trigonometric functions. A quick method of checking the equivalence of these results is to plot the functions or, even better, their difference.

. . . .

Try It

To further understand the different ways in which **simplify** and **normal** work, look at—and explain—the results of applying **simplify** and **normal** to the numerator and denominator of the trigonometric expression in Example 5-4.

The **normal** command does accept one optional argument, **expanded**. When this argument is specified, the factors in the numerator and denominator are expanded instead of being left in factored form.

EXAMPLE 5-5   ## A Different normal Form

Compare the results of calling **normal** for the expression $\dfrac{6}{x} + \dfrac{x-1}{(x+1)^2}$ both with and without the optional argument.

SOLUTION

The expression of interest in this example is

```
> EXPR := 6/x+(x-1)/(x+1)^2;
```

$$EXPR := \frac{6}{x} + \frac{x-1}{(x+1)^2}$$

The default normalized form for this expression is

```
> normal( EXPR );
```

$$\frac{7\,x^2 + 11\,x + 6}{x\,(x+1)^2}$$

Note that the numerator, which has no real roots, is expanded, but the denominator is still in its factored form. When the optional argument is specified, both numerator and denominator will be expanded:

```
> normal( EXPR, expanded );
```

$$\frac{7\,x^2 + 11\,x + 6}{x^3 + 2\,x^2 + x}$$

. . . .

5-3 USING factor

The **factor** command computes the factorization of a polynomial in one or more variables. In general, each factor has coefficients of the same type (for example, integer, floating point, or complex) as the coefficients of the original polynomial. You can force coefficients to be real (floating point) or complex valued by specifying **real** or **complex**, respectively, as a second (optional) argument to **factor**.

EXAMPLE 5-6

Three Factorizations of a Polynomial

Find the factorizations of $x^8 - 1$ with integer, real, and complex-valued coefficients. Find all solutions to $x^8 - 1 = 0$.

SOLUTION

Two real-valued roots of the $x^8 - 1$ can be seen on inspection: $x = 1$ and $x = -1$. Thus, $x - 1$ and $x + 1$ are both factors of this expression. The factorization of the remaining sixth-order polynomial is not obvious.

```
> EXPR := x^8-1;
```

$$EXPR := x^8 - 1$$

```
> EXPR2 := simplify( EXPR / (x-1) / (x+1) );
```

$$EXPR2 := x^6 + x^4 + x^2 + 1$$

The **factor** command, with a single argument, gives the factorization with integer coefficients:

```
> factor( EXPR );
```

$$(x - 1)(x + 1)(x^2 + 1) \, (x^4 + 1)$$

The corresponding factorization with floating-point coefficients is

```
> factor( EXPR, real );
```

$$(x + 1.)(x - 1.)(x^2 + 1.414213562x + 1.000000000)(x^2 + 1.)$$
$$(x^2 - 1.414213562x + 1.000000000)$$

Note that this factorization is approximate but does contain information that can be used to obtain an exact symbolic factorization in linear terms (see the Try It! exercise immediately following this example).

The factorization with floating-point coefficients is automatically returned if at least one of the coefficients of the argument is a floating-point number. For example, **evalf** can be used to change the integer 1 to a floating-point 1.:

```
> EXPRf := evalf( EXPR );
```

$$EXPRf := x^8 - 1.$$

and then factored to obtain

```
> factor( EXPRf );
```

$$(x + 1.)(x - 1.)(x^2 + 1.414213562x + 1.000000000)(x^2 + 1.)$$
$$(x^2 - 1.414213562x + 1.000000000)$$

The factorization with complex-valued floating-point coefficients is

```
> factor( EXPR, complex );
```

$$(x + 1.)(x + .7071067812 + .7071067812I)$$
$$(x + .7071067812 - .7071067812I)(x + 1. I) \, (x - 1. I)$$
$$(x - .7071067812 + .7071067812I)$$
$$(x - .7071067812 - .7071067812I)(x - 1.)$$

. . . .

Rational functions can also be factored using **factor**. In this case, Maple normalizes the expression and then factors the numerator and denominator before letting the automatic simplification algorithm produce the final result.

Try It   Find the factorization of $\dfrac{x^4 - y^4}{x^3 - y^3}$. Compare the results from **factor** with those from **simplify** and **normal**.

5-4 USING expand AND combine

The principal use of the **expand** command is to distribute products over sums, but **expand** can also be used for a variety of transformations involving trigonometric, exponential, logarithmic, and other functions. The inverse transformations are applied via the **combine** command.

EXAMPLE 5-7

Finding a Polynomial from its Roots

Find, in expanded form, a polynomial with roots -1, 2, $\dfrac{3}{2}$, and $\sqrt{5}$. Check your result in at least two different ways.

SOLUTION

Each root of the polynomial contributes one factor to the polynomial. This suggests

> `POLY := (x+1)*(x-2)*(x-3/2)*(x-sqrt(5));`

$$POLY := (x+1)(x-2)\left(x-\frac{3}{2}\right)(x-\sqrt{5})$$

When these four terms are multiplied, the expanded polynomial is

> `POLYe := expand( POLY );`

$$POLYe := x^4 - x^3\sqrt{5} - \frac{5}{2}x^3 + \frac{5}{2}x^2\sqrt{5} - \frac{1}{2}x^2 + \frac{1}{2}x\sqrt{5} + 3x - 3\sqrt{5}$$

There are a number of ways to check this result. For example,

> `factor( POLYe );`

$$\frac{1}{2}(x-2)(2x-3)(x+1)(x-\sqrt{5})$$

gives the polynomial in factored form. The leading coefficient is inserted to allow the factor corresponding to the root at $\dfrac{3}{2}$ to have integer coefficients. Note, however, the presence of $\sqrt{5}$, a non-integer, in the factorization (see Problem 4). Additional ways to verify this result include plotting the polynomial and looking at the x-intercepts, evaluating the polynomial at each of the intended roots, and solving the equation **POLYe = 0** for x.

. . . .

The summation formulas for sine and cosine,

$$\sin(u+v) = \sin(u)\cos(v) + \cos(u)\sin(v)$$
$$\cos(u+v) = \cos(u)\cos(v) - \sin(u)\sin(v),$$

are examples of the trigonometric transformations that **expand** attempts to apply to its argument.

EXAMPLE 5-8

Expanding Trigonometric Functions

Use **expand** to rewrite $\cos(2\theta + \phi)$ in terms of $\sin(\theta)$, $\cos(\theta)$, $\sin(\phi)$, and $\cos(\phi)$. Verify that **combine** converts the expanded expression back to its original form.

SOLUTION

The original expression is

> **EXPR := cos(2*theta + phi);**

$$EXPR := \cos(2\theta + \phi)$$

In this case, the result of the **expand** command is

> **EXPRe := expand(EXPR);**

$$EXPRe := 2\cos(\phi)\cos(\theta)^2 - \cos(\phi) - 2\sin(\phi)\sin(\theta)\cos(\theta)$$

Note that each term in this expression is a product of one or more of the desired terms.

To return to the original form, use **combine**:

> **EXPRc := combine(EXPRe);**

$$EXPRc := \cos(2\theta + \phi)$$

. . . .

Try It

To better understand how **combine** works, use **op** to extract the three terms from **EXPRe**, apply **combine** to each term, then reassemble the results. Is this the same as **EXPR**? Explain.

___

When the first argument to **expand** is a rational function, only the numerator is expanded.

EXAMPLE 5-9

An Algebraic Fraction

Rewrite $\dfrac{(x+y)\,(x-y)}{(x-2\,y)\,(x+3\,y)}$ as a rational function where both numerator and denominator are fully expanded.

SOLUTION

The expression to be studied in this example is

> **EXPR := (x+y)*(x-y)/(x-2*y)/(x+3*y);**

$$EXPR := \frac{(x+y)\,(x-y)}{(x-2\,y)\,(x+3\,y)}$$

It does not suffice to simply use **expand** since this command does not expand the denominator; moreover, **expand** separates the expression into two separate terms:

> **expand(EXPR);**

$$\frac{x^2}{(x-2\,y)\,(x+3\,y)} - \frac{y^2}{(x-2\,y)\,(x+3\,y)}$$

One means of forcing the expansion of the denominator is to explicitly divide the expanded forms of the numerator and denominator:

> **EXPRe2 := expand(numer(EXPR)) / expand(denom(EXPR));**

$$EXPRe2 := \frac{x^2 - y^2}{x^2 + x\,y - 6\,y^2}$$

An alternative means of obtaining this result is to use **normal** with its optional second argument (see the online help worksheet for **normal**):

> **normal(EXPR, expanded);**

$$\frac{x^2 - y^2}{x^2 + x\,y - 6\,y^2}$$

To further understand the expansion of rational functions, note the difference when **expand** is applied to the last result:

> **expand(EXPRe2);**

$$\frac{x^2}{x^2 + x\,y - 6\,y^2} - \frac{y^2}{x^2 + x\,y - 6\,y^2}$$

. . . .

In some situations, it is desirable to expand only selected parts of an expression. One way to prevent specific subexpressions from being expanded is to include these expressions as optional arguments to expand. The efficient use of this feature takes practice and experimentation, but the results can be worth the effort.

EXAMPLE 5-10   ## Simplifying Logarithmic and Exponential Expressions

Rewrite $\ln\!\left(\left(\dfrac{x}{x^2-1}\right)^{(2\,x+2)}\right) + (x+1)\,e^{(x+2)}$ in a simplified form involving logarithms of linear terms.

SOLUTION

The original expression is

```
> EXPR := ln((x/(x^2-1))^(2*x+2)) + (x+1)*exp(x+2);
```

$$EXPR := \ln\left(\left(\frac{x}{x^2-1}\right)^{(2x+2)}\right) + (x+1)\, e^{(x+2)}$$

Although the object of this example is to simplify the expression, **simplify** is not the appropriate Maple command for this job:

```
> simplify( EXPR );
```

$$\ln\left(\frac{\left(\frac{x^2}{(x-1)^2\,(x+1)^2}\right)^x x^2}{(x-1)^2\,(x+1)^2}\right) + e^{(x+2)}\,x + e^{(x+2)}$$

There are several ways to approach this problem. One is to first apply **factor** to the expression:

```
> EXPRf1 := factor( EXPR );
```

$$EXPRf1 := \ln\left(\left(\frac{x}{(x-1)\,(x+1)}\right)^{(2x+2)}\right) + e^{(x+2)}\,x + e^{(x+2)}$$

Note that although the exponential term is expanded into two pieces, the denominator of the logarithm is now factored. Now the **expand** command can be used to apply properties of the logarithm and exponential functions:

```
> EXPRe1 := expand( EXPRf1 );
```

$$EXPRe1 := 2\,x\ln(x) - 2\,x\ln(x-1) - 2\,x\ln(x+1) +$$
$$2\ln(x) - 2\ln(x-1) - 2\ln(x+1) + e^x e^2 x + e^x e^2$$

The common factor of $x+1$ in the coefficient of the logarithmic and exponential terms can be extracted using **factor**:

```
> EXPRf2 := factor( EXPRe1 );
```

$$EXPRf2 := (2\ln(x) - 2\ln(x-1) - 2\ln(x+1) + e^x e^2)(x+1)$$

The final step is to recombine the product of the two exponentials:

```
> EXPRc1 := combine( EXPRf2 );
```

$$EXPRc1 := (2\ln(x) - 2\ln(x-1) - 2\ln(x+1) + e^{(x+2)})(x+1)$$

One step in this process can be eliminated if $e^{(x+2)}$ is never replaced with $e^x e^2$. To prevent this part of the expansion, simply include each expression that should not be expanded as an optional argument to expand. In this case,

> `> EXPRe2 := expand( EXPRf1, x+2 );`

$$EXPRe2 := 2\,x\ln(x) - 2\,x\ln(x-1) - 2\,x\ln(x+1) + 2\,\ln(x) -$$
$$2\,\ln(x-1) - 2\,\ln(x+1) + e^{(x+2)}x + e^{(x+2)}$$

The final form can now be obtained by a single call to **factor**:

> `> EXPRc2 := factor( EXPRe2 );`

$$EXPRc2 := (2\ln(x) - 2\ln(x-1) - 2\ln(x+1) + e^{(x+2)})(x+1)$$

. . . .

You might be somewhat surprised that **combine** did not recombine the three logarithm terms into a single logarithm. This transformation was not used since (in real-valued calculus) $\ln(x) + \ln(y) = \ln(xy)$ only when x, $y > 0$, and Maple is working under the default assumption that all variables are complex-valued.

In addition to **op**, the **coeff** command is sometimes useful when you need to extract a specific subexpression from an expression. For example, **coeff(EXPR, exp(x+2));** returns $x + 1$, the coefficient of **exp(x+2)** in **EXPR**.

Try It

Repeat the previous example when x is assumed to be positive. Find conditions on x that allow all three logarithm terms to be combined into a single logarithm.

5-5 USING TYPES AND TYPE CONVERSION

Many of the descriptions of commands in this module have specifically mentioned the kinds of Maple objects (for example, set, list, expression, equation) that the commands expect as arguments. For example, recall that in Example 3-8 you saw that a set cannot be sorted—you can only sort a list. The original problem was conquered by converting the set to a list with the **convert** command.

EXAMPLE 5-11

Conversion between Lists and Sets

Use the same general methods as in Example 3-8 to construct sorted lists of the perfect squares and perfect cubes that do not exceed 1 million. Use these lists to construct the sorted list of integers through 1 million that are both perfect squares and perfect cubes.

SOLUTION

The only difference between this example and Example 3-8 is that the collections of perfect squares and cubes are explicitly requested as sorted lists. This change can be realized by changing the delimiters for the command that generates the expression sequence of elements.

```
> SQR := [ i^2 $ i = 1 .. 1000 ]:
> CUB := [ i^3 $ i = 1 .. 100 ]:
```

The elements in these lists are automatically sorted because the underlying expression sequence is created in increasing order. Since it is not possible to compute the intersecton of a list, these lists must be converted to sets before the intersection can be found.

```
> SCset := convert( SQR, set ) intersect convert( CUB, set );
```

$$SC := \{1, 64, 1000000, 262144, 117649, 46656, 15625,$$
$$729, 4096, 531441\}$$

The result is a set, so the elements can no longer be expected to be sorted. The sorted list is obtained with the **sort** command after converting the set to a list.

```
> SClist := sort( convert( SC, list ) );
```

$$SClist := [1, 64, 729, 4096, 15625, 46656, 117649,$$
$$262144, 531441, 1000000]$$

More generally, **convert** is used to change a Maple object from one type to another. There are approximately 100 built-in Maple types (see the help worksheet with key word TYPE). The most common types are **set**, **list**, **exprseq**, **name**, **integer**, **numeric**, **polynom**, **'+'**, and **'*'**. Once you realize that **exprseq** is an abbreviation for "expression sequence," and **polynom** is an abbreviation for "polynomial," the names are fairly self-explanatory.

The **type** command is a Boolean function (that is, it returns a value of **true** or **false**) used to test whether an expression has a specific type. A second command for testing types within an object is **hastype**. The key difference between **type** and **hastype** is that **type** returns true if the first argument is of the type specified in the second argument, and **hastype** returns **true** if the first argument has a subexpression of the specified type.

EXAMPLE 5-12

Comparison of type **and** hastype

Predict, and then verify, the results of applying **type** and **hastype** to the set

$\{1, \frac{2}{5}, a + b, [x\,y, \mathrm{e}^3], \sin(\theta)\}$ with types **set**, **list**, **'*'**, and **equation**.

SOLUTION

The set used in each test is

> ```
> SET := { 1, 2/5, a+b, [ x*y, exp(3) ], sin(theta) };
> ```

$$SET := \{\, 1, \frac{2}{5}, a+b, \sin(\theta), [\, x\, y, e^3\,]\, \}$$

Since **SET** is a set, both **type** and **hastype** should return the value **true**:

> ```
> type( SET, set ); hastype( SET, set );
> ```

$$true$$
$$true$$

Although the **type** command should return **false** for each of the other three tests, the set does contain a list and a multiplication so that the corresponding **hastype** commands should return **true**:

> ```
> type( SET, list ); hastype( SET, list );
> ```

$$false$$
$$true$$

> ```
> type( SET, '*' ); hastype( SET, '*' );
> ```

$$false$$
$$true$$

No element of **SET** contains an equation, so the final test should return false for both **type** and **hastype**:

> ```
> type( SET, equation ); hastype( SET, equation );
> ```

$$false$$
$$false$$

. . . .

Try It

The results in Example 5-12 show that three of the four possible combinations of **true** and **false** can be obtained when **type** and **hastype** are applied to the same arguments. Is it possible for **type** to return **true** and **hastype** to return **false** for the same arguments?

The third command used in typechecking is **whattype**. As the name suggests, the value returned by **whattype** is the type of its argument: in particular, the data type of the top-level data type as determined by the precedence ordering of the operators present in the argument. A full list of possible values returned by **whattype** can be found in the corresponding online help worksheet.

Even though almost all Maple types have fairly intuitive names, it can be important to pay close attention to some of the subtle differences between similarly named types. To illustrate, the **type** online help worksheet includes at least 15 types related to numbers: **numeric**, **positive**, **negative**, **nonneg**, **integer**, **posint**, **negint**, **nonnegint**, **even**, **odd**, **float**, **fraction**, **rational**, **constant**, and **realcons**.

EXAMPLE 5-13

Examination of Types of Numbers

A complicating fact about types is that a single expression can have more than one type. A careful reading of the online help explains that any expression of type **integer**, **fraction**, or **float** is also of type **numeric**. Determine which of the 15 types relating to numbers are associated with the values e^3, $\dfrac{3\pi}{2}$, $\cos\left(\dfrac{\pi}{2}\right)$, and $\ln(-\pi)$.

SOLUTION

To begin, assemble the relevant types in a single list (why is this preferable to a set?):

```
> numtypes := [numeric, positive, negative, nonneg,
>               integer, posint, negint, nonnegint, even, odd,
>               float, rational, fraction, constant, realcons ]:
```

The first expression to be tested is

```
> EXPR := exp(3);
```

$$EXPR := e^3$$

The full set of tests can be requested in a single command with the use of **seq**:

```
> seq( type( EXPR, TYPE ), TYPE = numtypes );
```

false, false, false, false, false, false, false, false, false, false, false,
false, false, true, true

The only types that return **true** are **constant** and **realcons**; **numeric**, **positive**, and **float** would apply only if the expression were reported as a floating-point number.

The second expression is

```
> EXPR := 3*Pi/2;
```

$$EXPR := \frac{3}{2}\pi$$

```
> seq( type( EXPR, TYPE ), TYPE = numtypes );
```

false, false, false, false, false, false, false, false, false, false, false,
false, false, true, true

Once again, only **constant** and **realcons** match; similar to the previous example, the symbolic π prevents the match with **numeric**, **positive**, and **float**:

> ```
> EXPR := cos( Pi/2 );
> ```

$$EXPR := 0$$

> ```
> seq( type( EXPR, TYPE ), TYPE = numtypes );
> ```

true, false, false, true, true, false, false, true, true, false, false,
true, false, true, true

Note that Maple automatically simplifies this expression to the integer 0; for this reason, this expression does have type **rational**.

> ```
> EXPR := ln( -Pi );
> ```

$$EXPR := \ln(-\pi)$$

What is the logarithm of a negative number? The results of this test should give you some ideas about this quantity:

> ```
> seq( type( EXPR, TYPE ), TYPE = numtypes );
> ```

false, false, false, false, false, false, false, false, false, false, false,
false, false, true, false

Without going into any great detail, this test suggests that $\ln(-\pi)$ is a complex-valued constant.

Try It Write a single Maple command that uses nested **seq** commands that carry out the 15-type tests for each of the 4 expressions. Be sure the results are well organized and easy to read.

Application 5   ## WATER QUALITY

Environmental engineers use criteria such as the level of dissolved oxygen, temperature, and indirect measures of the concentration of various organic and inorganic compounds to assess natural water quality. Fish in a natural stream typically require between 4 and 10 mg/L of dissolved oxygen (DO) to survive. Cooler water is denser and contains a higher concentration of dissolved oxygen than warmer water. Thus, the summer months place greater stress on aquatic life, and fish typically seek deeper (cooler) water during this season. We will use the Streeter–Phelps equation to model the DO sag curve for a stream exposed to pollutants that compete with aquatic life for the available DO.

Fundamentals

This problem involves two related, but separate, quantities: the amount of oxygen required to oxidize organic (waste) matter present in a body of water and the amount of DO available in the water to accomplish this task. The following sections explain the differential equation that governs each of these quantities. Explicit solutions are also provided. This analysis does not require knowledge of differential equations or calculus, except for the concept of rate of change. Several calculus-based examples and problems relating to this application can be found in Chapter 6 on the ftp site.

DO and BOD

Dissolved oxygen (DO) is the amount of molecular oxygen dissolved in water and is one of the most important criteria in determining natural water quality. DO also affects wastewater treatment processes. The water's carrying capacity for DO, also known as the DO *saturation level*, depends on the temperature of the water. DO saturation levels for different water temperatures can be measured with a DO meter; the values in Table 5-1 are typical. Note that cooler water contains a greater concentration of dissolved oxygen than warmer water.

Table 5-1   Relationship Between Temperature and DO Saturation Level (Measured Data)

Temperature (degrees C)	Dissolved Oxygen (mg/L)	Temperature (degrees C)	Dissolved Oxygen (mg/L)
0	14.6	16	9.9
1	14.2	17	9.7
2	13.9	18	9.5
3	13.5	19	9.3
4	13.1	20	9.1
5	12.8	21	8.9
6	12.5	22	8.7
7	12.1	23	8.6
8	11.8	24	8.4
9	11.6	25	8.3
10	11.3	26	8.1
11	11.0	27	8.0
12	10.8	28	7.8
13	10.5	29	7.7
14	10.3	30	7.6
15	10.1		

Biochemical oxygen demand (BOD) is the amount of oxygen required to oxidize organic matter that is biochemically present in water and is, therefore, an indirect measure of organic water contamination. The greater the BOD, the greater the oxygen depletion in a stream or lake. It is a measure of waste strength insofar as it measures the oxygen-consuming property of waste in terms of oxygen that is biologically consumed. The BOD (in mg/L) after t days is $\quad BOD_t = \dfrac{DO(0) - DO_t}{V_s/V_0}$,

where V_0 is the dilution bottle volume, V_s is the sample volume, and V_s/V_0 is the sample dilution. (For example, the sample dilution would be 30 to 1 for a 10-mL wastewater sample placed inside a 300-mL bottle filled with dilution water.) The BOD is usually measured under controlled conditions, such as a temperature T = 20°C and darkness (to prevent oxygen-producing algae). DO(0) and DO_t are the levels of dissolved oxygen in the sample bottle at the outset and after t days, respectively.

The rate of BOD consumption at each instant of time is proportional to the BOD remaining in the water supply at that time. That is, if L(t) is the remaining BOD at time t, then L′(t) = $-k_d$L(t), where the deoxygenation rate, $k_d > 0$, (with units of 1/time) depends on a number of factors, including the number and type of microorganisms and the water temperature. Since k_d is positive, the organic contaminants decay exponentially with time. You will learn how to use Maple to solve this differential equation in Chapter 6 (see the ftp site). The solution is L(t) = $L_0 e^{-k_d t}$, where L_0 is the BOD remaining at the outset ($t = 0$). Let y(t) = L_0 − L(t) denote the amount of oxygen consumed through time t in mg/L. Assuming this is the only process affecting the oxygen content of the water sample, the total amount of oxygen will be conserved. That is, for all $t \geq 0$, y(t) + L(t) = L_0 = constant. Note that the constant can be determined, by measurement, at any instant of time; $L_0 = BOD_u$, the ultimate BOD, is the total amount of BOD (waste) available for consumption (at $t = 0$). It is also the total amount of consumed oxygen when all the waste is depleted (when $t \to \infty$).

For example, suppose the EPA measures the 5-day and 10-day oxygen consumption as BOD_5 = 220 mg/L and BOD_{10} = 290 mg/L, respectively. The consumed oxygen, BOD remaining, and ultimate BOD corresponding to this data are displayed in Figure 5-1 as a function of the number of days t.

The DO Sag Curve and the Streeter–Phelps Equation

The Streeter–Phelps equation accurately models the amount of DO in a stream after wastewater is discharged into it. This model follows the pollutant downstream as it travels at the stream velocity. When a pollutant is introduced into a water source, the DO typically decreases to a minimum before gradually recovering to the saturation level. The plot of the DO as a function of time is called the DO sag curve. There are two competing processes in this interaction: reaeration and deoxygenation.

Figure 5-1
$y(t)$, $L(t)$ and BOD_u (in
mg/L) *vs.* time (in days);
BOD_u = 323 mg/L and
k_d = 0.228/day

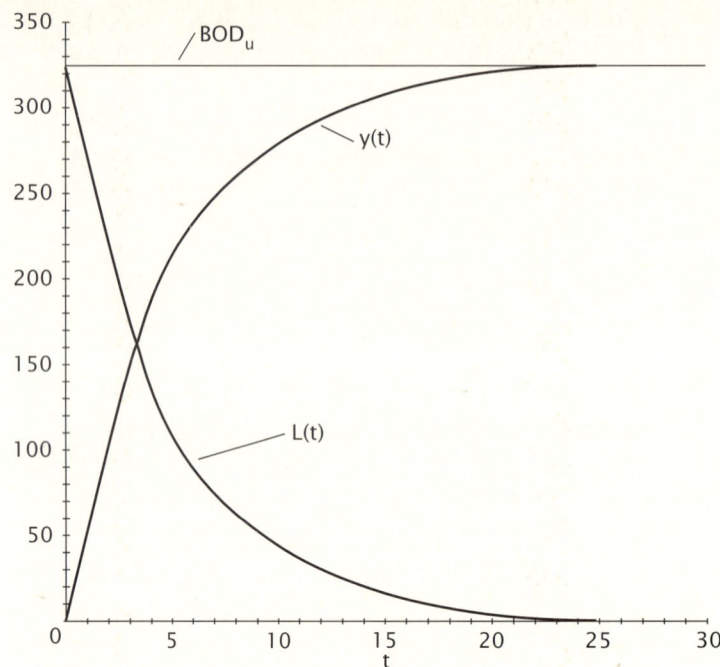

Figure 5-1
$y(t)$, $L(t)$ and BOD_u (in mg/L) *vs.* time (in days); BOD_u = 323 mg/L and k_d = 0.228/day

Reaeration *adds* molecular oxygen to the stream from the atmosphere (up to the saturation point); deoxygenation *depletes* the oxygen. Only the biochemically degradable microorganisms responsible for BOD are considered in the present analysis.

Let k_d denote the deoxygenation rate (per day), k_r the reaeration rate (per day), $D(t)$ the oxygen deficit in the stream (the difference between the saturation and the actual DO level), and $L(t)$ the stream BOD remaining at time t. (Note that k_d and $L(t)$ are the same quantities discussed in the first part of this application.) The Streeter–Phelps model states that the rate of change of the stream oxygen deficit, $D(t)$, increases in direct proportion to the stream BOD remaining, with proportionality constant given by the deoxygenation rate k_d, since the BOD is an indirect measure of organic water contamination itself. On the other hand, the rate of change of the oxygen deficit D decreases in direct proportion to the deficit at time t, with proportionality constant given by the reaeration rate k_r. The Streeter–Phelps equation for the oxygen deficit, D, represents both of these interacting processes: $D'(t) = k_d L(t) - k_r D(t)$. The solution to the Streeter–Phelps equation is

$$D(t) = \frac{k_d \, BOD_u \, (e^{-k_d t} - e^{-k_r t})}{k_r - k_d} + D_0 \, e^{-k_r t}$$

where D_0 is the oxygen deficit when the pollutant first enters the stream ($t = 0$). Suppose $k_d = 0.4$/day, $k_r = 2.0$/day, $BOD_u = 54.8$ mg/L, and the initial DO level is 2.2 mg/L at a stream temperature of 21°C. The DO sag curve (Figure 5-2) displays the dissolved oxygen, that is, the difference of DO_{sat} and $D(t)$, as a function of t: $DO(t) = DO_{sat} - D(t)$.

The minimum of the DO sag curve, which occurs at the sag time, is the time when the oxygen deficit $D(t)$ is greatest (minimum DO) and represents the time of greatest stress to fish in the stream. Since the pollutant is flowing downstream at the stream velocity, it is necessary to identify both when and where the minimum is attained.

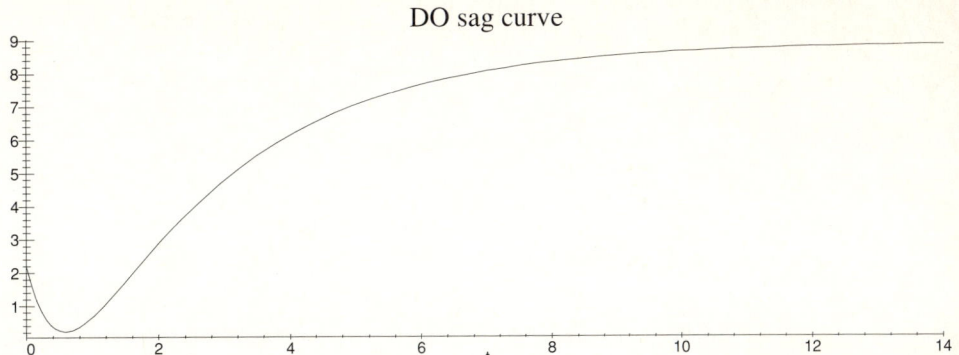

Figure 5-2
DO sag curve with
DO_{sat} = 8.9 mg/L,
BOD_u = 54.8 mg/L,
k_d = 0.4/day,
k_r = 2.0/day

■ 1. Define the problem

Use the solution to the Streeter–Phelps model to calculate both the minimum dissolved oxygen value in a stream and how far downstream the spill travels before attaining this minimum. Assume the stream velocity is 20 km/day.

■ 2. Gather information

This problem seems, at first, to be rather complicated because of the number of parameters, functions, and equations that are involved. You will see, however, that the problem is not too difficult if the information is carefully organized.

The initial value problems for the remaining BOD, L(t), and DO deficit, D(t), form the foundation of the problem:

$$L'(t) = -k_d L(t), \qquad\qquad L(0) = L_0$$

$$D'(t) = k_d L(t) - k_r D(t), \qquad D(0) = D_0.$$

The parameters in this model are the deoxygenation and reaeration rates k_d and k_r, and the initial ($t = 0$) conditions are the ultimate BOD, $L_0 = BOD_u$, and the oxygen deficit level, D_0, when the pollutant is added to the water. The solutions to these initial value problems are, provided $k_d \neq k_r$:

$$L(t) = BOD_u\, e^{-k_d t}$$

$$D(t) = \frac{k_d\, BOD_u\, (e^{-k_d t} - e^{k_r t})}{k_r - k_d} + D_0\, e^{-k_r t}$$

Other quantities of interest in the analysis include the amount of oxygen consumed by the waste organisms, y(t) = L_0 – L(t), and the dissolved oxygen level in the stream, $DO(t) = DO_{sat} - D(t)$, where DO_{sat} is the temperature-dependent saturated DO level (see Table 5-1). The water temperature is denoted by $T\,(°C)$ and the stream velocity by $V\,(km/day)$.

The parameters needed for the current analysis are

$$k_d = 0.4/\text{day}, \quad k_r = 2.0/\text{day}, \quad BOD_u = 54.8 \text{ mg/L},$$
$$DO_0 = 2.2 \text{ mg/L}, \quad T = 21°C, \quad V = 20 \text{ km/day}$$

◢ 3. Generate and evaluate potential solutions

To begin the analysis, the parameter values can be used to graph the DO sag curve. An estimate of the sag time can be obtained from a graph of the DO sag curve. This graph will be created by finding a general expression for DO(t) and then substituting all relevant parameter values.

The general solution to the Streeter–Phelps equation provides an explicit formula for the oxygen deficit D(t):

```
> restart;
> Deqn := DD = kd/(kr-kd)*BODu*(exp(-kd*t) - exp(-kr*t)) +
            Do*exp(-kr*t);
```

$$Deqn := DD = \frac{kd\ BODu\ (e^{(-kd\,t)} - e^{(-kr\,t)})}{kr - kd} + Do\ e^{(-kr\,t)}$$

The corresponding equation for the DO level in the water is obtained from the fact that the total dissolved oxygen is conserved:

```
> DOconserv := DO + DD = DOsat;
```

$$DOconserv := DO + DD = DOsat$$

Note that deficit D(t) is denoted by **DD** since **D** is the name of a built-in Maple command.

Now, the formula for dissolved oxygen (DO) is found to be

```
> DOeqn := op( solve( DOconserv, { DO } ) );
```

$$DOeqn := DO = -DD + DOsat$$

```
> DOeqn := subs( Deqn, DOeqn );
```

$$DOeqn := DO = -\frac{kd\ BODu\ (e^{(-kd\,t)} - e^{(-kr\,t)})}{kr - kd} - Do\ e^{(-kr\,t)} + DOsat$$

```
> DOeqn1 := collect( ", { exp(-kd*t), exp(-kr*t) } );
```

$$DOeqn1 := DO = -\frac{kd\ BODu\ e^{(-kd\,t)}}{kr - kd} + \left(\frac{kd\ BODu}{kr - kd} - Do \right) e^{(-kr\,t)} + DOsat$$

The specific parameters for this problem are:

```
> DOvals := [ kd=0.4, kr=2.0, DOo=2.2, T=21, BODu = 54.8 ];
```

$$DOvals := [kd = .4, kr = 2.0, DOo = 2.2, T = 21, BODu = 54.8]$$

Before the DO sag curve can be plotted, it is necessary to determine both the initial oxygen deficit, D_0, and the DO saturation, DO_{sat}. The saturation level is, from Table 5-1, with T = 21°C, DO_{sat} = 8.9 mg/L; the corresponding deficit is $D_0 = DO_{sat} - DO_0 = 8.9 - 2.2 = 6.7$ mg/L:

```
> DOvals := [ op(DOvals), Do=6.7, DOsat=8.9 ];
```

$$DOvals := [kd = .4, kr = 2.0, DOo = 2.2, T = 21,$$
$$BODu = 54.8, Do = 6.7, DOsat = 8.9]$$

```
> DOeqn2 := subs( DOvals, DOeqn1 );
```

$$DOeqn2 := DO = -13.70000000\,\mathrm{e}^{(-.4\,t)} + 7.00000000\,\mathrm{e}^{(-2.0\,t)} + 8.9$$

```
> plot( rhs( DOeqn2 ), t=0..14, title='DO sag curve' );
```

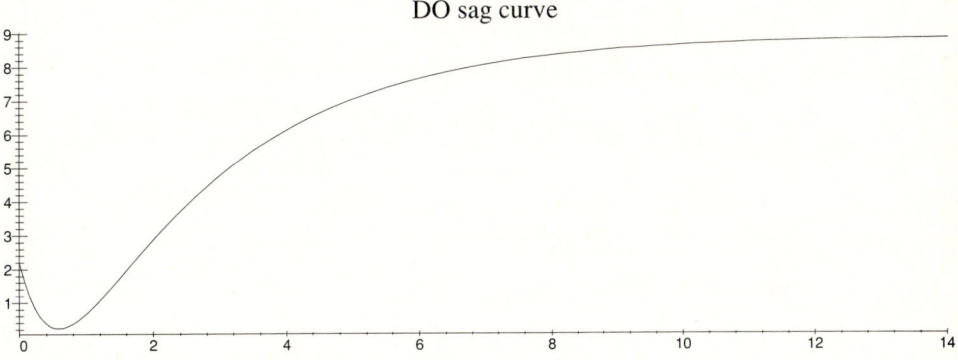

DO sag curve

The sag time is estimated, using Maple's graphics interface, to be t_{crit} = 0.575 days, or 13.8 hours. Since the stream velocity is 20 km/day, the minimum DO level, which is DO = 0.224 mg/L, occurs 11.5 km downstream from the contamination injection point.

Observe that the initial DO level for this problem is almost 50 percent lower than the minimum level needed to support fish life. The DO level finally reaches the level of 4 mg/L after 2.525 days, when the spill is 50.5 km downstream. Beyond this point, the DO level increases but does not exceed the saturation level (8.9 mg/L).

4. Refine and implement a solution

The estimates in Step 3 can be improved by zooming in on the portion of the graph near the sag time. A more sophisticated approach is to observe that the minimum DO level occurs when the oxygen deficit is at its maximum value. Moreover, the maximum deficit occurs at the instant in time where the tangent line to the graph of D(t) is horizontal. This condition is equivalent to finding the time when the right-hand side of the Streeter–Phelps equation is zero.

The Streeter–Phelps equation is

```
> StrPhlps := diff( DD(t), t ) = kd*L(t) - kr*DD(t);
```

$$StrPhlps := \frac{\partial}{\partial t} DD(t) = kd\, L(t) - kr\, DD(t)$$

The **diff** command is used to compute derivatives in Maple. It is used here only to make the equation look like a differential equation; see Section 6-2 for an introduction to calculus using Maple.

In general, the slope of the tangent line to the graph of the oxygen deficit curve is the rate of change of the oxygen deficit at that instant in time. The tangent line is horizontal when the slope is zero. This leads to the equation

```
> DDhoriz_eqn := rhs(StrPhlps) = 0;
```

$$DDhoriz\_eqn := kd\, L(t) - kr\, DD(t) = 0$$

The solution to the Streeter–Phelps equation has already been entered:

```
> Deqn;
```

$$DD = \frac{kd\, BODu\,(e^{-kd\, t} - e^{-kr\, t})}{kr - kd} + Do\, e^{-kr\, t}$$

The solution of the initial value problem for the remaining BOD is

```
> Leqn := L = BODu * exp( -kd*t );
```

$$Leqn := L = BODu\, e^{(-kd\, t)}$$

Substitution of these equations into the right-hand side of the Streeter–Phelps equation requires two steps. The first, is replacement, for example, of L(t) with L, and the second is substitution of the exact solutions:

```
> sag_eqn := subs( [ L(t)=L, DD(t)=DD ], DDhoriz_eqn );
```

$$sag\_eqn := kd\, L - kr\, DD = 0$$

```
> sag_eqn := subs( [ Leqn, Deqn ], sag_eqn );
```

$$sag\_eqn := kd\, BODu\, e^{(-kd\, t)} - kr\left(\frac{kd\, BODu\,(e^{(-kd\, t)} - e^{(-kr\, t)})}{kr - kd} + Do\, e^{(-kr\, t)} \right) = 0$$

In this form, the equation looks rather forbidding. Notice, however, that the equation is actually only a linear combination of two exponential functions:

```
> sag_eqn := collect( sag_eqn, { exp(-kd*t), exp(-kr*t) } );
```

$$sag\_eqn := \left(kd\,BODu - \frac{kr\,kd\,BODu}{kr - kd} \right) e^{(-kd\,t)} - kr \left(-\frac{kd\,BODu}{kr - kd} + Do \right) e^{(-kr\,t)} = 0$$

The coefficient of the exponential that decays with rate k_d can be simplified further. To prevent the destruction of the previous modifications, it is advisable to extract this coefficient,

```
> COEFF := coeff( lhs(sag_eqn), exp(-kd*t) );
```

$$COEFF := kd\,BODu - \frac{kr\,kd\,BODu}{kr - kd}$$

perform the necessary simplifications,

```
> normal( COEFF );
```

$$\frac{kd^2\,BODu}{-kr + kd}$$

and then reinsert the coefficient into the equation for the sag time

```
> sag_eqn := subs( COEFF=", sag_eqn );
```

$$\frac{kd^2\,BODu\,e^{-kd\,t}}{-kr + kd} - kr \left(-\frac{kd\,BODu}{kr - kd} + Do \right) e^{-kr\,t} = 0$$

The sag time predicted by this equation is

```
> tcrit := solve( sag_eqn, { t } );
```

$$tcrit := \left\{ t = -\frac{\ln\left(\dfrac{kr\,(kd\,BODu - Do\,kr + Do\,kd)}{kd^2\,BODu} \right)}{-kr + kd} \right\}$$

◢ **5. Verify and test the solution**

To conclude this analysis, the solutions found in Steps 3 and 4 will be compared. Using the same set of parameters as in Step 3, the sag time is found to be

```
> sag_time := subs( DOvals, tcrit );
```

$$sag\_time := \{t = .6250000000\,\ln(2.554744526)\}$$

```
> sag_time := evalf( op(sag_time), 3 );
```

$$sag\_time := t = .585$$

This is almost 0.01 days (almost 15 minutes) later, and 0.2 km further downstream, than was predicted from the graph. The discrepancy between the floating-point and graphical results can be reduced, as mentioned in Step 4, by zooming in on the region around the sag time. However, an accurate approximation is difficult to obtain because the curve is essentially horizontal for a relatively long period of time in the neighborhood of the critical time. The result ($t_{crit} = 0.585$ days) obtained analytically in Step 4 does not suffer from this difficulty; it gives the true sag time, within the accuracy of the model.

The formula for the DO level as a function of time was found in Step 3:

```
> DOeqn2;
```

$$DO = -13.70000000 \, e^{(-.4\,t)} + 7.00000000 \, e^{(-2.0\,t)} + 8.9$$

The minimum level of DO occurs at the sag time:

```
> evalf( subs( sag_time, DOeqn2 ), 3);
```

$$DO = .27$$

This level is still well below the minimum needed to support fish life in the stream. The time when the DO level returns to the lower limit of 4 mg/L is

```
> fsolve( subs( DO=4, DOeqn2 ), t );
```

$$2.548663995$$

This is noticeably closer to the estimate obtained in Step 3.

What If

Suppose a wastewater treatment plant is constructed at a certain location along the stream. After the treatment water is mixed with the upstream water, we find that the water temperature just downstream (after mixing) is $T = 26°C$, $DO = 6.9$ mg/L, $BOD_u = 15.2$ mg/L, and the stream velocity is 20 km/day. The deoxygenation and reaeration rates are the same as the earlier upstream values. Plot the new DO sag curve and identify the critical time, t_{crit}, where the DO is at a minimum, and find this minimum value. How far downstream from the treatment plant does this minimum occur? Are there any portions of the stream downstream from the treatment plant where fish cannot survive?

SUMMARY

In this chapter, you learned to use the Maple commands **simplify**, **collect**, **normal**, **factor**, **expand**, and **combine** to perform detailed manipulation of expressions, sets, lists, and other Maple objects. Also important in these types of operations are the use of **type**, **hastype**, and **whattype** to identify the type of an object, **convert** to change objects into a desired type, and the **assume** facility and side relations to provide additional information about a Maple object.

You also worked through an application involving wastewater treatment analysis to solve a problem similar to that which an environmental engineer would encounter. Differential equations were used to model the biochemical oxygen demand and oxygen deficit in a stream. The solution to the Streeter–Phelps equation was used to plot the DO sag curve and to determine, both graphically and analytically, the time when and where the oxygen content of the stream was at a minimum. In the next chapter you will learn how to use Maple to solve these differential equations, and you will also learn how to apply Maple in several different branches of mathematics.

Key Words

expanded form	relatively prime
factored form	side relations
irreducible polynomial	simplification rule
rational expression	

Maple Commands

assume
about
coeff
convert
diff
expand
combine
factor
is
normal,
   optional argument: **expanded**
select
remove
simplify,
   optional arguments: **assume=**,
symbolic type, **hastype**,
whattype

types:
set, **list**, **exprseq**, **name**,
integer, **numeric**, **polynom**, **'+'**,
'*', **numeric**, **positive**,
negative, **nonneg**, **integer**,
posint, **negint**, **nonnegint**, **even**,
odd, **float**, **fraction**, **rational**,
constant, and **realcons**

References

1. Ray, B.Y., *Environmental Engineering* (Chapter 8), Barton: PWS Publishers, 1995.
2. Kennedy, M.S. and Bell, J.M., "The Effects of Advanced Wastewater Treatment on River Water Quality," *J. Water Pollution Control Federation*, Vol. 58, No. 12, 1986, pp. 1139–44.
3. Nicolaides, R. and Walkington, N., *Maple: A Comprehensive Introduction*, Cambridge: Cambridge University Press, 1996.
4. Sincero, A.P. and Sincero, G.A., *Environmental Engineering: A Design Approach*, Cliffs, N.J.: Prentice Hall, 1996, pp. 37–43 and 187–198.

Problems

1. How does Maple simplify $\sqrt{z^2}$ when z is complex? real? positive? negative? Explain all results. Since this expression involves only a single name, is there any difference between using the **assume** command and the **assume=** optional argument to **simplify**? (Be sure to look at the online help for any functions that you have not seen previously.)

2. Determine conditions on z so that $\sqrt{e^z} = e^{z/2}$. *Hint:* An equivalent form of this question is when is $\sqrt{e^z} - e^{z/2} = 0$?

3. Symbolic simplification should not be overused. To see some of the potential pitfalls, consider the expression $\left((-2)^p\right)^{1/p}$.

   (a) Compute the value of this expression for $p = -5, -4, -3, -2, -1, -2/3, -1/3, 0, 1/2, 1, 5/4, 3/2, 7/4, 2, 3, 4, 5$.

   (b) What does Maple simplify this expression to when p is complex? positive? negative? even? odd?

   (c) How does Maple simplify this expression when the **symbolic** option is used in **simplify**?

   (d) For what values of p are the answers in parts (b) and (c) consistent?

4. Although it is a well-known fact from algebra that $x^2 - 2 = (x - \sqrt{2})(x + \sqrt{2})$, this result is not obtained from **factor(x^2-2);**. The explanation for this can be seen in the fact that the factorizations returned by **factor** generally have integer coefficients. To include integer multiples of one or more specific nonintegers, called extensions to the field of integers, include these numbers as a set as the second argument to factor. For example, **factor(x^2-2, { 2^(1/2) });** returns the expected factorization for $x^2 - 2$. Irrational numbers that appear in the polynomial are automatically included in the set of extensions (see Example 5-7).

   Find appropriate sets of extensions that yield the full factorization of

   (a) $x^2 + 4x - 41$ and (b) $x^3 - \dfrac{5\,x^2}{2} - 5\,x + \dfrac{3}{2}$.

5. (a) Use the factorization of $x^n - 1$ to obtain the roots of $x^n = 1$ with as much accuracy as possible for each $n = 1, 2, 3, 4, 5, 6, 7, 8$.

   (b) Use the **complexplot** command, from the **plots** package, to plot all solutions to $x^n - 1$, for $n = 1, 2, 3, 4, 5, 6, 7, 8$.

   (c) Compare your results in part (a) with the results obtained by using **solve** to find the solutions to $x^n = 1$.

6. Find all values of the parameter a for which the functions
   $f(x) = x^2 + ax + 26$ and $g(x) = x^4 + 6x^3 - 17x^2 - 78x - 56$ have at least one common root.

7. The expression **EXPRel** in Example 5-10 is not a valid simplification of **EXPR** for all real and complex values of x. Find values of x that give different values when inserted into **EXPR** and **EXPRel**. Find the general conditions on x that guarantee that the two expressions are equivalent.

8. The polynomial with roots -1, 2, $\dfrac{3}{2}$, and $\sqrt{5}$ was found in Example 5-7. Find, also in expanded form, the polynomial with the same roots but with coefficients that sum to 1.

9. Example 5-13 presents a number of questions that are worth pursuing. Foremost is the question about the logarithm of a negative number. One way to get more insight into this question is to look at a floating-point approximation to $\ln(-\pi)$. Although this can be done using **evalf**, find a way to achieve the same result using **convert**.

10. The values tested in Example 5-13 matched different combinations of the 15 types related to numeric objects. Is it possible to find one number that matches all 15 types? If not, what is the highest number of matches that can be made with a single number?

11. (a) It is well known that the sine of all integer multiples of π is zero: $\sin(n\pi) = 0$ for all integers n. Add assumptions to the name n so that Maple automatically simplifies $\sin(n\pi)$ to zero. What is the value of $\cos(n\pi)$ for any integer n?

    (b) Consider the expression $\sin\left(\dfrac{n\pi}{2}\right)\cos\left(\dfrac{n\pi}{2}\right)$. Use the **combine** command to simplify this expression. Now, add the assumption that n is an integer. How is this extra information reflected in the original and combined expressions? (Explain any differences in the results.)

    (c) Another lesson related to **assume** is that assumptions should be imposed only after all other simplifications have been completed. For example, compare the results of applying **combine** to $\sin\left(\dfrac{n\pi}{2}\right)\cos\left(\dfrac{n\pi}{2}\right)$ with and without the assumption that n is an integer.

12. Properties and types are closely related. One difference is that some properties can be specified in a convenient mathematical form: for example, **assume(z>0);**. Just as the **type** command is used to test types, the **is** command is used to test if a Maple object has a specific property (see the **assume** help worksheet). The value returned by **is** will be **true** (if the property follows from the previous assumptions), **false** (if the property is not always consistent with the assumptions), or, **FAIL** (if Maple was not able to determine whether the property is true or false).

    (a) Verify that, when Maple knows $x > 2$, **is(x^2 + 2*x + 3 > 2);** returns the value **false** and **is(x^2+2*x+3 >= 2);** returns the value **true**.

(b) Determine appropriate properties to impose on z so that
`is( ln(z^2+1) > 0 );` returns the value **true**. How can the
assumptions on z be relaxed so that `is( ln(z^2+1) >= 0 );`
evaluates as **true**?

13. (a) Determine at what time 99.5% of BOD_u is attained in Figure 5-1.

(b) Find the exact time when the BOD reaches 99.5% of the ultimate
BOD for general values of the reaction rate, k_d, and the ultimate
BOD, BOD_u. Explain how this time depends on both parameters.

(c) Repeat b) for any threshold (not just 99.5% of BOD_u). That is, determine the time until a sample with reaction rate k_d reaches p% of
the ultimate BOD.

14. (a) Find, and plot, the linear function that best fits (in the least
squares sense) the DO vs. temperature data in Table 5-1.

(b) Find, and plot, the exponential function that best fits this data.

(c) How do the two fits compare? Which looks to be the better fit? For
each fit, compute the sum of the squares of the difference between
the absolute error between the measured and predicted values.
What does this say about the quality of the two fits?

Chapters 6 and 7 can be found at
<http://www.awl.com/cseng/toolkit/modules/map>.

6 Advanced Engineering Computations

INTRODUCTION

Maple can be used in many different branches of mathematics including algebra, calculus, combinatorics, financial mathematics, graph theory, linear algebra, logic, number theory, and optimization. The first five chapters of this module showed you how to use Maple to solve problems requiring mainly algebra and trigonometry. In this chapter, you will see some of the ways Maple can be used to solve problems that involve complex analysis, calculus, and (briefly) differential equations. The application in this chapter shows how an electrical engineer uses nodal analysis to investigate some of the characteristics of a low-pass electrical filter. Recall that the bandwidth application in Chapter 3 also involved the analysis of a filter. Even though there are similarities in the ways engineers discuss both types of filters, the discussion in this chapter is quite different from the one presented in Chapter 3.

Although you will learn many useful techniques for solving mathematical problems in this chapter, it is not comprehensive in its presentation of Maple's mathematical capabilities. Some of the examples and Try It! exercises refer to problems encountered earlier in the module. For example, the solution to the Streeter–Phelps equation that was provided in Chapter 5 will now be found and verified using Maple. References to the earlier discussions are provided; if you do not recall the details of the problem, please take a few minutes to refresh your memory.

Several dozen packages contain collections of additional Maple commands. The **plots** and **student** packages have been introduced previously. Here, the **student** package will be used in the discussion of calculus (Section 6-2) and the **DEtools** package will be used to produce plots relating to differential equations (Section 6-3). Other packages likely to be of interest to many engineers include the **linalg** package for linear algebra, the **numtheory** package for number theory, and the **inttrans** package for integral transformation such as the Laplace and Fourier transform. The full list of Maple packages can be found on the online help worksheet with keyword **index,package**.

7

Introduction to Maple Programming

The first six chapters presented techniques for direct interaction with Maple: enter a command, receive a response. While this is sufficient for many situations, there are times when repeatedly stepping through complicated multi-command sequences is inconvenient and inefficient. The main topic of this chapter is the use of Maple as a programming language to create user-defined commands, including loops, conditionals, input arguments, error handling, and return values.

The simplicity of Maple programs, more properly called *procedures*, is derived from the fact that the programming is done using standard Maple commands. The power of the Maple programming language is evidenced by the fact that almost all Maple commands are implemented in the Maple programming language. Moreover, as you will learn in section 7.1, the definitions can be viewed, and even modified, by the user.

Some of the examples and problems present a final look at examples and applications from previous chapters. The application introduced in this chapter investigates some of the analytical techniques used by chemical engineers in the design of reactors for the mixing of solvents and solutes.

Index